Stem Cell Biology and Regenerative Medicine

Series Editor
Kursad Turksen
Ottawa, Ontario, Canada

Our understanding of stem cells has grown rapidly over the last decade. While the apparently tremendous therapeutic potential of stem cells has not yet been realized, their routine use in regeneration and restoration of tissue and organ function is greatly anticipated. To this end, many investigators continue to push the boundaries in areas such as the reprogramming, the stem cell niche, nanotechnology, biomimetics and 3D bioprinting, to name just a few. The objective of the volumes in the Stem Cell Biology and Regenerative Medicine series is to capture and consolidate these developments in a timely way. Each volume is thought-provoking in identifying problems, offering solutions, and providing ideas to excite further innovation in the stem cell and regenerative medicine fields.

More information about this series at http://www.springer.com/series/7896

Kursad Turksen
Editor

Bioprinting in Regenerative Medicine

Humana Press

Editor
Kursad Turksen
Ottawa
Ontario
Canada

ISSN 2196-8985 ISSN 2196-8993 (electronic)
Stem Cell Biology and Regenerative Medicine
ISBN 978-3-319-21385-9 ISBN 978-3-319-21386-6 (eBook)
DOI 10.1007/978-3-319-21386-6

Library of Congress Control Number: 2015947956

Springer Cham Heidelberg New York Dordrecht London

Printed on acid-free paper

Humana Press is a brand of Springer
Springer International Publishing AG Switzerland is part of Springer Science+Business Media (www.springer.com)

Preface

While our understanding of stem cells has advanced tremendously since the seminal work of Till and McCulloch in the 1960s, our ability to translate this knowledge to the clinical setting for tissue repair and restoration of function following injury and disease lags behind. Significant efforts have been made to mimic tissue and organ architecture in cell cultures, with some encouraging but generally small-scale results, but achieving true 3D complex organ structures has been elusive. Thus, the development of 3D bioprinting technologies for repair or even replacement of tissues and organs has been an exciting new approach to the problem.

To provide a comprehensive and state of the art summary of 3D printing for tissue engineering, I have attempted to recruit several outstanding groups who are actively involved in moving this field forward. I am extremely grateful for their willingness to contribute chapters that cover the fundamentals of the technology and its utility with stem and other cells to demonstrate how it may accelerate the ultimate goal of restoring function of damaged tissues and organs.

I would like to acknowledge Aleta Kalkstein for helping me get this project off the ground and her encouragement during its completion.

A special thank you goes to Emily Janakiram who was instrumental in facilitating not only timely completion of the project but its completion with the highest Springer standards.

Ottawa, 2015 Kursad Turksen

Contents

Contributors

Sujata Kumari Bhatia School of Engineering and Applied Sciences, Harvard University, Cambridge, MA, USA

Kathleen Chen Department of Biomedical Engineering, Rensselaer Polytechnic Institute, Troy, NY, USA

David T. Corr Department of Biomedical Engineering, Rensselaer Polytechnic Institute, Troy, NY, USA

Xiaofeng Cui School of Chemistry, Chemical Engineering and Life Sciences, Wuhan University of Technology, Wuhan, Hubei, China

Center for Biotechnology and Interdisciplinary Studies, Rensselaer Polytechnic Institute, Troy, NY, USA

Stemorgan Therapeutics, Albany, NY, USA

Guohao Dai Department of Biomedical Engineering, Rensselaer Polytechnic Institute, Troy, NY, USA

Andrew Dias Department of Biomedical Engineering, Rensselaer Polytechnic Institute, Troy, NY, USA

James B. Hoying Chief, Division of Cardiovascular Therapeutics, Cardiovascular Innovation Institute, University of Louisville, Louisville, KY, USA

Xavier Intes Department of Biomedical Engineering, Rensselaer Polytechnic Institute, Troy, NY, USA

Esmaiel Jabbari Biomimetic Materials and Tissue Engineering Laboratory, Department of Chemical Engineering, University of South Carolina, Columbia, SC, USA

Chemical and Biomedical Engineering, Swearingen Engineering Center, University of South Carolina, Columbia, SC, USA

Bahattin Koc Department of Manufacturing and Industrial Engineering, Sabanci University, Istanbul, Turkey

Can Kucukgul Department of Manufacturing and Industrial Engineering, Sabanci University, Istanbul, Turkey

Vivian K. Lee Department of Biomedical Engineering, Rensselaer Polytechnic Institute, Troy, NY, USA

Seyedsina Moeinzadeh Biomimetic Materials and Tissue Engineering Laboratory, Department of Chemical Engineering, University of South Carolina, Columbia, SC, USA

Dileep Daniel Monie Faculty of Arts and Sciences, Harvard University, Cambridge, MA, USA

S. Burce Ozler Department of Manufacturing and Industrial Engineering,, Sabanci University, Istanbul, Turkey

Mehmet S. Ozturk Department of Biomedical Engineering, Rensselaer Polytechnic Institute, Troy, NY, USA

Seyed Ramin Pajoum Shariati Biomimetic Materials and Tissue Engineering Laboratory, Department of Chemical Engineering, University of South Carolina, Columbia, SC, USA

Brad Tricomi Department of Biomedical Engineering, Rensselaer Polytechnic Institute, Troy, NY, USA

Stuart K. Williams Bioficial Organs Program, Cardiovascular Innovation Institute, University of Louisville, Louisville, KY, USA

About the Editor

Kursad Turksen received his Ph.D. from the University of Toronto in the area of osteoprogenitor biology and cell selection methodologies. He then did postdoctoral training at the Howard Hughes Medical Institute, University of Chicago, studying epidermal biology through use of genetically-altered mouse models. He joined the Ottawa Hospital Research Institute [previously known as the Loeb Research Institute] where he rose to the position of Senior Scientist in the Division of Regenerative Medicine. His research interests focus on stem cell biology, with a particular interest in the Claudin family of tight junction proteins and their role in epidermal lineage commitment and progression during development in health and disease.

Dr. Turksen is currently Editor-in-Chief of Springer's journal *Stem Cell Reviews and Reports* in addition to serving as series editor for *Stem Cell Biology and Regenerative Medicine.*

Bioinks for Bioprinting

Stuart K. Williams and James B. Hoying

1 Introduction

1.1 Elements of 3D Bioprinting

The history of additive manufacturing that has evolved into the current applications known as 3D-printing is based on the basic technique of layer by layer assembly of structures. The original materials for additive manufacturing were primarily soft woods that could be quickly shaped into each layers form (J. E. Blanther, "Manufacture of Contour Relief Maps ", US Patent #473,901, 1892). Several traditional Rapid Prototyping techniques have been exploited and adapted for generating scaffolds, such as fused deposition modeling (FDM) [1–3], three-dimensional printing (3DP) [4, 5] selective laser sintering (SLS) [3, 6–10], and stereolithography apparatus (SLA) [3, 11, 12]. The emergence of 3D—Bioprinting has resulted in the development of numerous materials for the assembly of biological structures. This chapter will review the materials that have been selected for various forms of bioprinting.

3D Bioprinting represents multiple components and elements that work synergistically starting with an image or structure that the operator wishes to print,

S. K. Williams (✉)
Bioficial Organs Program, Cardiovascular Innovation Institute, University of Louisville, Louisville, KY 40202, USA
e-mail: cvregen@gmail.com

J. B. Hoying
Chief, Division of Cardiovascular Therapeutics, Cardiovascular Innovation Institute, University of Louisville, Louisville, KY 40202, USA
e-mail: jay.hoying@louisville.edu

K. Turksen (ed.), *Bioprinting in Regenerative Medicine,* Stem Cell Biology and Regenerative Medicine, DOI 10.1007/978-3-319-21386-6_1

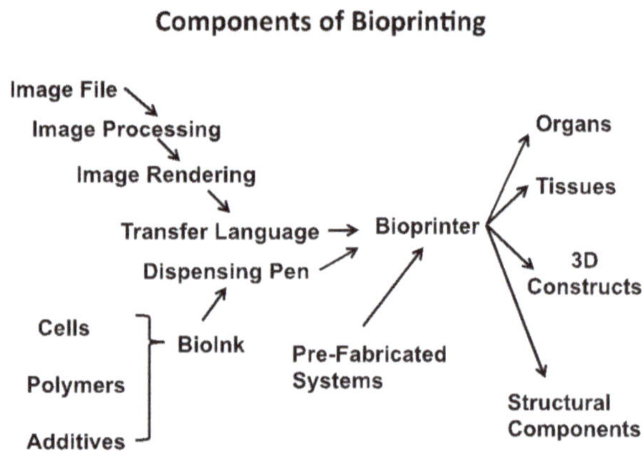

Fig. 1 Basic components of bioprinting

software and hardware solutions and the resulting printed structures. The basic elements of 3D Bioprinting are identified in Fig. 1.

The Bioinks that have been developed for 3D printing utilize several types of dispensing systems for the delivery of the material. Two of these dispensing systems are illustrated in Fig. 2. Dr. Tom Boland first described the use of a modified ink jet printer to print biological materials [13]. These studies utilized a commercial ink jet printer and ink jet cartridge. The cartridge was cleared of ink and the ink replaced with a suspension of cells in a gel. The basic structure of a piezo electric ink jet printer cartridge is illustrated in Fig. 2. There are also thermal ink-jet printer designs. The dispensing system used for 3D Bioprinting have evolved and now

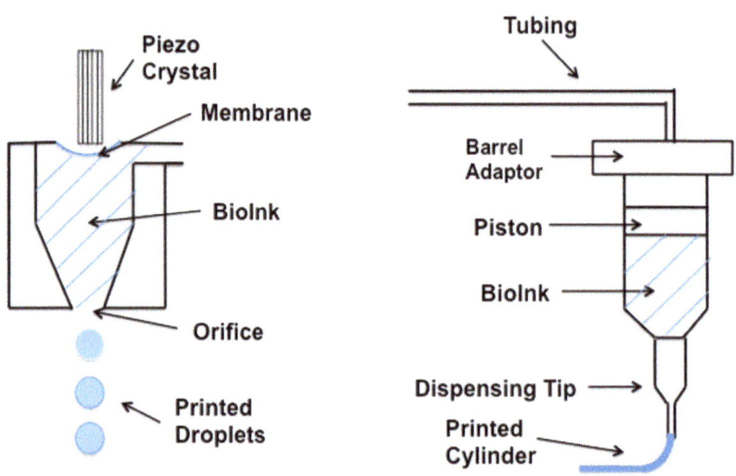

Fig. 2 Bioink dispensing pens. *Left* is the ink jet printing design where regular ink is replaced with bioinks. *Right* is the Time/pressure dispensing system used in many direct write bioprinters

include several types of dispensing methods that include; Ink jet printing [13–17], direct write time-pressure dispensing [18–29], Laser assisted printing [30–34].

The time-pressure needle dispensing systems is one of the oldest and most developed dispensing technologies, which normally consists of a syringe (barrel) containing a material that is directly attached to the dispensing tip (Fig. 2). Air pressure or a mechanical plunger (positive displacement) is employed to force the material through a dispensing needle in a time-controlled manner. Pressure is removed to stop material flow at the end of the dispensing. The amount of material dispensed is proportional to the amount and duration of the applied pressure. This technology is widely incorporated into bioprinters because of its low cost, easy setup and material flexibility. However, the technology can be prone to variability due to several factors including clogging of dispenser tips, a build-up of material at the dispensing tip sometimes referred to as “snow plowing” and a narrow range of pressure and time to permit rapid polymer flow rates while maintaining cell viability.

The time-pressure dispensing system has become one of the most broadly used systems in 3D Bioprinters where the bioink being dispensed is composed of a liquid biomaterial. Time-pressure dispensing systems are also called Direct-write printing systems [20, 22], Nozzle printing [35, 36] and Free-Form Fabrication [37]. The basic structure of a common time-pressure dispensing system is illustrated in Fig. 3. There are several manufacturers of time-pressure dispensing systems including GPD Global (Grand Junction, CO), Nordson (East Providence RI) and nScrypt (Orlando, FL) and OK International (Garden Grove, CA). The characteristics of time-pressure dispensing systems has been reviewed extensively with extensive evaluation of the effect of dispensing tip geometry, viscosity and pressure/vacuum characteristics of different systems [35, 38–40]. Most of these studies involve extrusion nozzles with single lumens. Of note, Ozbolat and colleagues have developed a novel dual lumen nozzle with the capability to form tubular structures at the time of dispensing [39]. This technology will be critical for the formation, using 3D bioprinting approaches, of small tubular structures (e.g. blood vessels).

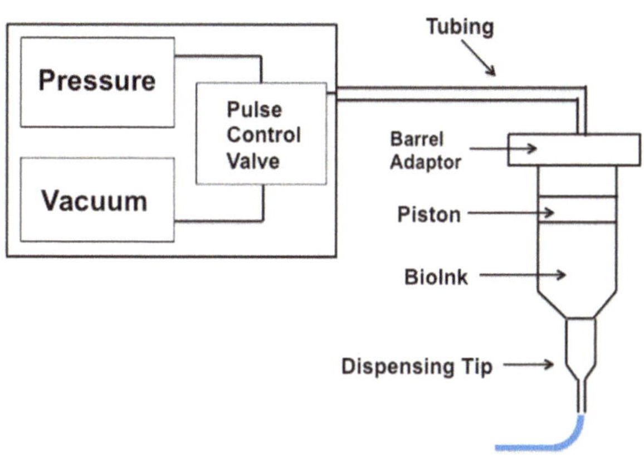

Fig. 3 A time/pressure dispensing system

1.2 Bioinks—Gels and Solutions

The origins of bioinks for 3D bioprinting can be found in the development of 3D cell culture techniques. The culture of cells in 3D is based on the realization that, in the body, cells exist primarily in a three-dimensional environment. This environment supports both cell-cell communication as well as cell-extracellular matrix communication. The culture of cells in two-dimensions often results in the loss of cellular differentiation and loss of cell function. The earliest studies evaluating the culture of cells as three-dimensional constructs, often defined as organ culture, were performed by Charles Lindbergh in the 1930s [41–44]. Understanding that the function of cells in vitro and in vivo was influenced by the extracellular matrix was realized in the early 1970s [45–48] and quickly led to the use of three-dimensional in vitro culture of cells [49–57]. The earliest gels used for three-dimensional tissue culture are based on natural substances including fibrin gels and collagen [56–59]. Interest in using additive manufacturing techniques for the study of cell function was first explored using 3D printing to create complex material scaffolds followed by addition of cells to the scaffold material [60]. This approach has been followed by methods to create complex geometries with incorporation of specific cell binding sites [61, 62]. All of these technologies have provided a foundation for the technology known as cellular bioprinting, the process whereby cells and a supporting gel are simultaneously extruded through a printing pen to create complex tissue constructs. The first material used to bioprint cells with viability of the cell maintained was agar [63]. Since these earliest studies numerous materials have been used as the solution component of bioinks. These materials are summarized in Table 1.

Table 1 Common Materials Used in Bioinks and Mechanism of Gel Formation

Compound	Mechanism gel formation	Chemical structure
Agar	Thermal	Polysaccharide
Collagen	Spontaneous gelation/ photoinitiation	Protein
Alginate	Ionic	Polysaccharide
PLGA-g-PEG	Thermal	Poly(lactic-*co*-glycolic acid)
PEGDMA	Thermal/chemical	Poly(ethylene glycol) dimethacrylate
Pluronic	Thermal	Poly(ethylene glycol)-poly(propylene glycol)-poly(ethylene glycol)
Agarose	Thermal	Polysaccharide
Carageenen	Thermal	Polysaccharide
Fibrin	Spontaneous gelation	Protein
Elastin	Photoinitiation	Protein
Silk	Photoinitiation	Protein
Chitosan	Chemical	Polysaccharide
Hyaluronic acid	Chemical	Glycosoaminoglycan
NIPAAM	Thermal	N-isopropyl acrylamide/N-t-butyl acrylamide copolymer

1.3 Materials Used as Bioinks in Bioprinting

Agar is a polysaccharide first described in Japan in the seventeenth century by an innkeeper, Minoya Tarozaemon, who noted the gelation of seaweed based soup after the soup froze during the night. Two centuries later agar was isolated from algae. The major components of agar are agarose (major components D-galactose and 3,6-anhydro-L-galactopyranose) and agaropectin (major components d-galactose and 3,6-anhydro-l-galactose). Agar exhibits the ability to transition between a solution at high temperatures and forms a gel at lower temperatures [64], a process often called hysteresis. Agar has been used extensively in 3D bioprinting [65–72] and was the first material used in a bioink by Tom Boland during his pioneering studies of bioprinting [63]. The typical method for agar –based bioprinting is warming of the agar solution until melted at ~85 °C and the transition to a gel at ~40 °C. This temperature allows bacterial cell bioprinting as these cells will survive at temperatures >40 °C, however, mammalian cells are not viable at this elevated temperature.

Agarose is a polysaccharide that is purified from agar. It has been used in many cell related applications due to its biocompatibility. One major use has been the ability to make gels of different structures and s the basis of what is known as the under-agarose cellular migration assay [73–75]. Agarose is available in numerous configurations including a low melting point version that remains as a liquid at 37 °C and solidifies at room temperature [76]. Agarose has also been used in many bioprinting applications to form scaffolds and molds for subsequent bioprinting of viable cells [67, 69, 71, 77]. The temperature characteristics permit the printing of a scaffold or phantom that is subsequently surrounded by a gel forming material that is not temperature sensitive. The temperature of the construct is then raised causing the agarose to revert to its solution form. The agarose is subsequently washed from the construct leaving channels. These channels can then be treated with cells to form blood vessel like structures [21, 25, 78–80].

Alginate is a anionic polysaccharide found in the cell walls of brown algae and seaweed. It was first identified by its chemical structure as alginic acid, and is also called align. When alginic acid is bound to water, capable of absorbing 300 times its weight in water, it quickly forms a gum like material. For this reason one of its original and still current uses is to form dental impressions [81], leading to its current use as a carrier of stem cells for dental applicaitons [82].

Alginic acid was first identified by the British chemist E. C. C. Stanford as an extract from brown algae and patented the method for purification in 1881 [64]. Since that time alginate has been used in multiple applications including as a bandage [83, 84], implantable material [85, 86], an encapsulation material [87–95], a gel for 3D culture of cells [95–97] and finally as a gel forming material for 3D bioprinting [24, 37, 40, 62, 69, 88, 95, 98].

Alginate is a linear copolymer with homopolymeric blocks of (1–4)-linked β-D-monnuronate (M) and its C-5 epimer α-L-guluronate (G) residues. The monomers can appear in homopolymeric blocks of consecutive G-residues (G-blocks), consecutive M-residues (M-blocks) or alternating M and G-residues (MG-blocks). When considering the use of alginate for bioprinting the commercial sources often provide basic information regarding the molecular weight of the alginate and the M/G ratio. Some of the commercial sources of alginate include Sigma Chemical Company, FMC Corporation, KIMIKA Corporation, Nacalai Tesque Inc. and Novamatrix Inc. Alginates are available from these sources in different sizes and MG ratios. Ueno and Oda have evaluated the biological activities of alginates [99], and report differences in cytokine and nitric oxide production by cells exposed to these alginates. When considering the use of alginate for bioprinting the presence of endotoxin should be considered as this lipopolysaccharide can activate cells. Selimoglu and colleagues have recently described methods to purify alginate to remove endotoxin [100]. There are filtration systems effective for the removal of endotoxin from alginate available from CUNO Corporation. Alginate manufactured according to GMP regulations is available from NovaMatrix®, a business unit of FMC BioPolymer.

As already described, nearly all commercial alginates today are produced from marine brown algae or brown seaweed. Alginates with more complex compositions can be isolated from bacteria such as Vibrio spledidus [101] and Azotobacter vinelandii [102] which produces polymers containing more complex structures. Production by fermentation therefore is technically possible but is not economically feasible at the moment. There has been significant progress in the understanding of alginate biosynthesis over the last 10 years. The fact that the alginate molecule enzymatically undergoes a post-polymerization modification with respect to chemical composition and sequence opens up the possibility for in vitro modification and tailoring of commercially available alginates.

For 3D bioprinting applications most investigators take advantage of the ability of alginate to rapidly form gels when contacting solutions of $CaCl_2$ [64, 84, 103, 104]. The relationship between alginate concentration and $CaCl_2$ concentration toward the formation of calcium alginate gels continues to be studied [105, 106]. Alginate exhibits the ability to bind divalent ions in the following relative affinity: $Mg<Ca<Sr<Ba$. For bioprinting applications Ca is most commonly used. Recent studies have established the original predications of Rees and colleagues that alginate gelation results from the gelation is due to the association of sequences of guluronate residues within alginate [106]. Further research is necessary to establish how different formulations of alginate form gels with divalent cations and how these formulations may result in alginate gels that exhibit variability in porosity and stiffness. For bioprinting the viscosity of alginate is dependent on concentration (higher viscosity at higher concentration), temperature (higher viscosity at lower temperature) and molecular weight (higher viscosity at higher molecular weight). Alginate also exhibits decreased viscosity at high shear rates. For bioprinting applications the rapid extrusion of alginate through delivery pens will result in reduced viscosity.

Chitosan is a linear polysaccharide composed of randomly distributed N-acetyl-D-glucosamine (acetylated unit) and β-(1-4)-linked D-glucosamine (deacetylated unit). It is a natural occurring substances that is obtained by treating shrimp and other crustacean shells with the sodium hydroxide. Chitosan has been used extensively in the agriculture industry due to its anti-fungal activitiy and as a food additive. More recently it has been used in many tissue engineering applications [92,

107–133]. The methods for 3-D printing using chitosan are generally based on extrusion of solutions of solubilized chitosan followed by rapid chemical crosslinking usually involving NaOH. [134] Chitosan based scaffolds, created using photopolymerization, have been used as a framework or lattice for seeding cells [135].

Carageenan is another seaweed derived polysaccharide that differs from other algae and seaweed polysaccharides as it contains sulfur groups and exhibits a more helical structure in its native form. It has been used as a food additive for decades based on the relative ease of preparation and the ability to rapidly form gels. Carrageenan has been used in several tissue engineering applications including 3D printing [92, 132, 136–140]. One unique aspect of carrageenan is the ability to control the porosity of scaffolds during the gelation process to create scaffolds with characteristics that support cellular invasion [141].

Collagen is an abundant, naturally occurring protein in the body that has been actively used to create cell containing gels to understand the effect of a three-dimensional environment on cell function [45, 52, 142–146]. Collagen is the most abundant protein in the body accounting for more than 20 % of the total protein content. There are now more than 28 structurally different types of collagen that have been identified in mammals. The most common types of collagen are designated Types I, II, III, IV and V. The general location within the body that these collagens are found is provided below:

- Type I: "Mature" collagen found in skin, tendon, vascular ligature, organs, bone. This collagen is found in most scars after wounding.
- Type II: The main component of cartilage
- Type III: "Young" collagen found throughout the interstitium in young individuals. This collagen is replaced by the stiffer collagen type I during maturation.
- Type IV: This collagen forms basal lamina, the cell-secreted layer of the basement membrane.
- Type V: This collagen is found on many cell surfaces.

Collagen type I is the most commonly used collagen type for formation of gels for in vitro studies of cell function. This collagen is typically isolated from either rat tail [142], calf skin [147] or human placenta [148–150], based on modifications of the original methods of Gallop [151]. The current preparation of collagen I involves a limited proteolytic treatment of raw material (e.g. rat tail tendon, calf skin or human placenta followed by cycles of polymerization and re-solubilization using alternating cold acidic buffer to solubilize the type I collagen and warm neutral pH buffer to allow polymerization of the collagen into fibrils [52, 152]. There are numerous commercial sources of collagen type I or alternatively the collagen can be prepared from raw materials as needed.

The other major types of collagen (Types II through V) are found in lesser amount in the body and their isolation and purification is more complex then type I collagen. Just as type I collagen interacts with cells causing cellular differentiation these other collagens are also important for cellular function and differentiation. For example endothelial cells exhibit a different phenotype when plated onto type

I collagen as compared to type IV collagen, where type IV collagen stimulates the formation of endothelial tubes and type I collagen causes endothelial cells to stay in a more proliferative non-differentiated phenotype [45].

There has been extensive use of collagen type I to perform 3D bioprinting of cells [11, 18–22, 25, 38, 92, 153–161]. This due, in part, to the ability of collagen type I monomers to undergo fibrillar collagen formation when the temperature and pH of the fibril solution is raised to 37 °C and the pH is adjusted to neutrality. The process of collagen fibril formation results in the gelation of the collagen and the collagen gel will maintain its structure based on the concentration of collagen in the initial solution. Although this collagen solution to gel methodology is used extensively in cell biology including, now, 3D bioprinting, the mechanisms underlying the formation of fibrils has only recently received extensive study [162]. Clearly other constituents in the collagen type I solutions (e.g. ions, peptides and proteins) have a significant effect on not only the gelation of the collagen but also the physical characteristics of the collagen gel (e.g. density, stiffness). Moreover, other constituents in the extracellular matrix, specifically members of the laminin family of extracellular matrix proteins have profound effect on cellular function including the process of angiogenesis and neovascularization [163–166].

Gelatin has an extensive history of use in the assembly of three-dimensional gels for tissue engineering [50, 78, 87, 88, 110, 111, 129, 167–185] and has also been used as a component of bioinks [78, 88, 92, 186]. Gelatin is collagen that has been subjected to complete (usually thermal) hydrolysis. This hydrolyzed collagen has been used extensively to coat tissue culture plastic to support cell adherence during culture [149, 171] indicating that gelatin maintains the ability to interact with cellular membrane proteins such as integrins [185].

Hyaluronic acid also called hyaluronan or simply by the acronym HA is an anionic glycosaminoglycan distributed throughout the body predominantly in connective, epithelial, and neural tissues. It is unique among glycosaminoglycans in that it is nonsulfated, and can be very large, with a molecular weight greater than 1 million. HA has been used extensively in cell biology and tissue engineering studies [64, 78, 92, 125, 157, 160, 184, 186–205]. Hyaluronic acid has been reported to have biological effects on cellular phenotype including cell proliferation and cell migratory activity [157, 204, 206, 207]. Many 3D bioprinting studies that utilize hyaluronic acid involve first, additive manufacturing techniques to create a scaffold and the subsequent treatment of the scaffold with gels containing hyaluronic acid [157, 208]. The direct bioprinting of cells within hyalurinc acid solutions has been accomplished and has been used successfully to create heart valves [209]. The formation of stable structures using hyaluronic acid often uses photocrosslinking or chemical crosslinking during the printing process [78, 186].

Silk is a protein fiber composed mainly of fibroin and is produced by many insect larvae during the formation of cocoons. The best-known silk is obtained from the cocoons of the larvae of the mulberry silkworm *Bombyx mori*. [92, 179, 210–216]. Silk protein fibers have been used extensively in tissue engineering applications that include the use of silk to create scaffolds for subsequent cell transplantation [92, 179, 210–220]. Silk protein fibers have seen limited use in bioprinting to date but initial studies have illustrated the potential applications of silk in direct write printing of cell laden constructs [92, 213, 216, 221, 222]. Work by Schacht and colleagues [222] have established the ability to bioprint cells in silk bioinks without the need for crosslinking.

Fibrin is a fibrous protein that is formed naturally in the body during the process of coagulation. Fibrin (also called coagulation factor Ia) is formed following the action of the protease, thrombin, on fibrinogen (coagulation factor I). Fibrous fibrin exhibits the ability to undergo spontaneous gel formation. These fibrin gels have been used extensively in tissue engineering and cell biology studies, especially to study the effect of three-dimensional environment on cell function [11, 92, 158, 195, 204, 223–241]. Fibrinogen and fibrin have been used in 3D printing of cellular constructs utilizing the spontaneous gelation characteristics of fibrin [11, 62, 92, 158, 238, 242]. Several advantages for the use of fibrin in bioprinting, beyond just spontaneous gelation characteristics, include the natural occurrence of this process in the body and therefore biocompatibility of fibrin as an implant material, and the established effects of fibrin on cell function. Cells, and especially endothelium, exhibit the ability to interact with and undergo extensive neovascularization in fibrin gels [56, 57, 241, 243–249]. Beyond acting as a simple encapsulating gel, fibrin exhibits numerous biological effects on cellular function [226, 235, 250].

Elastin is also a naturally occurring extracellular matrix protein. Inherent in its name is the ability of elastin to undergo transition between a coiled and elongated form and the ability to provide elasticity to tissues such as skin and large caliber blood vessels. Elastin scaffolds can be obtained from tissues samples using both enzymatic and chemical de-cellularization and can subsequently be used as scaffolds for 3D printing [195, 251–253]. Very often elastin is co-printed with collagen or other materials that provide spontaneous or chemically augment cross-linking and gelation [254].

Thermosensitive Hydrogels There are numerous hydrogels used in 3D bioprinting applications that exhibit a temperature sensitive transition between a solution and gel form. This is often called a volume phase transition and is described by the ability to undergo transition from solution to gel at either a lower critical transition temperature (LCST) or upper critical transition temperature (UCST). Unlike chemical crosslinking that most often leads to permanent, non-reversible formation of gels, thermosensitive hydrogels have the ability to undergo repetitive gel to solution to gel transitions. LCST hydrogels exhibit the ability to undergo a solution to gel transition at increasing temperatures. Thus these polymers are gels at lower temperatures and solutions at higher temperatures. An example of an LCST is **Poly(N-isopropylacrylamide) or PNIPAM**. The transition temperature is 32 °C and thus PNIPAM can be bioprinted as a gel at temperatures below 32 °C and then converted to a solution at temperatures above 32 °C. Another example of LCST hydrogels is poly(N-isopropylacrylamide) or PIPAAm. The NIPAM and PIPAAM hydrogels have seen extensive use in tissue engineering where sheets of cells can be grown on the thermosensitive polymer gel at higher temperature and then the cell sheet can be released by simply lowering the temperature below 32 °C [255–257].

PEG Hydrogels are composed of synthetic crosslinked hydrogels of poly(ethylene) glycol (PEG) and have been used in numerous tissue engineering applications. Polyethylene glycol is a polyether compound with many industrial applications including use in medical applications. PEG is also known as polyethylene oxide (PEO) or polyoxyethylene (POE), depending on the molecular weight. There have been extensive studies on PEG biocompatibility with general agreement that these compounds exhibit biocompatibility. **PEGylation** is the process of attaching the strands of the polymer PEG to molecules, most typically proteins, drugs and antibodies, with the result of increased solubility and reduced immunogenicity. PEGylation also results in changes in receptor binding and can alter therapeutic effects. Two commonly used PEGylated hydrogels used in bioprinting are **PLGA-g-PEG and PEGDMA. PLGA-g-PEG,** poly(D, L-lactic acid-co-glycolic acid) with polyethylene glycol side chains, exhibits a lower critical transition temperature (LCST) [258–260] and bioprinting is performed at lower temperatures. **PEGDMA**, poly(ethylene glycol) dimethacrylate, is a water soluble co-monomer used in the manufacturing of plastics and as a crosslinking agent. PEGDMA has been used in many tissue engineering applications [158, 261–263] and has been used selectively in bioprinting applications [159]

Poloxamers—(The most common poloxamer is Pluronic) has been used extensively in 3D printing both as a scaffold or mold to support subsequent bioprinted structures [22, 238, 264–266] as well as direct bioprinting of cells with poloxamers [267]. A recent review by Alaxender and colleagues provides an in depth evaluation of the chemistry and gel forming mechanisms of poloxamers [264]. The word "poloxamer" was first used by the inventor, Irving Schmolka, who received the patent for these materials in 1973. The predominant commercial source of poloxamers is the BASF corporation and they are more commonly known by the trade names Synperonics [268],Pluronics [268] or Kolliphor [269]. Poloxamers are triblock copolymers composed of a central hydrophobic chain of *polyoxypropylene* (poly(propylene oxide/PPO)) flanked by two hydrophilic chains of polyoxyethylene (poly(ethylene oxide/PEO)).

When the PEO/PPO (2:1) is immersed into the aqueous solvents, they form micellar structures above critical micellar concentration. The poloxamers are quite soluble in aqueous, polar and non-polar organic solvents and are quite stable. The pluronic triblock copolymers are available in various grades differing in molecular weights and physical forms. Depending upon the physical designation for the grades are assigned, as F for flakes, P for paste, L for liquid.

As an example, the most commonly use poloxamer in 3D printing has been Pluronic F-127. The F designates this poloxamer has the physical characteristics of a flake/powder at room temperature. It is a poloxamer and has the following chemical structure:

$$\mathrm{H(OCH2CH2)_x(OCH2\overset{\overset{CH_3}{|}}{C}H)_y\,(OCH2CH2)_z\,OH}$$

The numerical designation, 127, provides information regarding the approximate molecular weight (first two digits multiplied by 300), and the last digit × 10 gives the percentage polyoxyethylene content. Therefore F127 is a flake at room temperature, has a polyoxypropylene molecular mass of 36,000 g/mol and a 7% polyoxyethylene content.

Pluronic F127 has a characteristic property of thermoreversible gelation that is advantageous in 3D printing. The thermoreversible characteristics of Pluronic 127 are observed in aqueous solutions in a concentration range of 20–30% w/w. Pluronic 127 is a liquid when refrigerated (4–5 °C) and turns into gel form when brought to room temperature (> 16 °C). The gel formed is reversible on cooling and this can be repeated (e.g. solution-gel-solution) multiple times. The gel formation occurs only when concentration is above critical micellar concentration. This thermoreversible gelation property is useful in the various drug delivery systems such as oral, ocular, nasal, topical, dental, and other biomedical fields. For 3D bioprinting the thermoreversible gelation properties of Pluronic F127 have been utilized to form tissue constructs of specific dimensions. Chang et al. [21–23]. It is also possible to print structures at 37 °C, fill and surround the printed structures structures with a non-thermosensitive gel, and by lowering the temperature of construct create channels or voids hen the Pluronic reverts to a solution form [22].

There are numerous examples where combinations of polymers, with different material characteristics, are added together in bioinks to provide multiple functionality. This is most often seen where a non gelling material (e.g. extracellular matrix—non collagen I) is added to a material that will form a gel under specific conditions [90, 92, 130, 192, 215, 240, 267, 270–278]. In certain cases once gelation has occurred the gel forming polymer can be removed. Examples are Pluronic and alginate used to create solute containing gels, followed by either thermal reversal of gelation [22, 192, 265–267, 279, 280], or alginate dissolution using EDTA [281].

1.4 Additives to Bioinks to Influence Cellular Behavior

Following the bioprinting of cells there is typically an incubation period, either in vitro or in vivo, that results in changes in cellular function include proliferation, migration, differentiation, apoptosis and self assemble. These cellular activities are regulated by multiple factors that include soluble factors such as growth factors and cytokines, the extracellular matrix and numerous small molecules. One of the most complex additives that has been used in bioinks is matrigel, a complex mixture of both growth factors, cytokines and extracellular matrix proteins [157, 166, 175, 274, 277, 282–296]. The addition of matrigel to bioinks is an appropriate first step in identifying whether a complex mixture of components/additives can support desired cellular function in the printed structures [157, 175]. The role of specific factors will require purified cytokines, growth factors, matrix proteins and peptides.

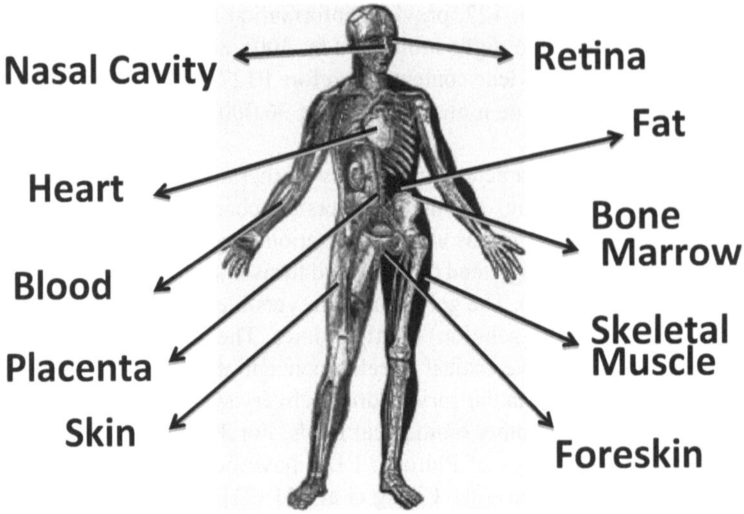

Fig. 4 Sources of cells for regenerative medicine

1.5 Cells for Bioprinting

A third and critical component of bioinks are the cells that populate tissues throughout the body. The initial studies of cell bioprinting utilized robust mammalian cell lines that can be maintained in 2D tissue culture with minimal media requirements. Numerous commercial sources are available for cell lines (e.g. Lonza; RoosterBio; American Type Culture Collection; Life Technologies and BioTime) that represent the major cell types found in the body. Figure 4 illustrates the major sources of cells used in tissue engineering and regenerative medicine.

As bioprinting moves from the laboratory to the clinic sources of clinical grade cells will be necessary to support the assembly of different constructs. Moreover, point-of-care isolation of a patients own cells for bioprinting may be necessary to avoid the use of immunosuppressive drugs. Point of care systems for the isolation of regenerative and stem cells have reached the stage f clinical trials and may soon be available as a source of cells for clinical bioprinting [297–300]

1.6 Chemical and Photosensitive Cross Linking of Solutions and Gels

The initial bioprinting of polymers into a specifically designed shape (e.g. spheroid [95, 301, 302], rod [20, 28, 303, 304], tube [39]), requires that the material either maintains its shape due to inherent viscosity or undergoes some form of stabilization or crosslinking to maintain shape. The crosslinking of biomaterials and natural substances to form a stable structure has been studied extensively and includes both chemical crosslinking [305–311] and photocrosslinking. Nimni and colleagues studied numerous aldehydes and other chemical crosslinkers and concluded, for collagen, glutaraldehyde provides the most biocompatible and stable end-product. Due to concerns regarding the possible toxicity of aldehydes, alternative chemical cross linking agents have been explored [312].

Other chemical cross linkers used to make tissue engineered constructs or bioprinted constructs include **Genipin**, **1-Ethyl-3-(3 dimethylaminopropyl) carbodiimide/N-hydroxysuccinimide (EDC/NHS** [313], **Tetrahedral polyethylene glycol** [78]. **Genipin** is a natural product extracted from the gardenia fruit, which overcomes the toxicity inherent in most commonly used synthetic cross-linkers [181, 276, 312]. Genipin can be utilized to crosslink functional amine groups present in natural tissues and biomaterials with very minimal cytotoxic effects, compared with glutaraldehyde [181, 276, 312, 314, 315]. Genipin cross-linked materials acquire a deep blue color following treatment. The utilization of genipin to crosslink natural biocompatible polymers, such as chitosan and gelatin, and to form biodegradable hydrogels has the potential to produce novel scaffolds for various tissue engineering and bioprinting applications. Carbodiimide containing fixatives have been used as an alternative to glutaraldehyde for fixation of tissues [316–318].

Photoactivated crosslinking of molecules to form stable structures has been extensively studied [159, 319–333]. This includes studies evaluating the formation of covalent linkage between molecules, especially related to surface modification of biomaterials [321, 333–336]. The use of photoactivated chemistries in bioprinting has seen increased interest [11, 30, 158, 159, 186, 238, 261, 327, 337] and is based on extensive foundational research on photoactivated chemistries. A large amount of the literature in photoinitiated cross linking of materials is based on the dental field where composite materials and resins are often set in place with light activated chemistries [176, 320, 338–340].

One of the most widely used UV photoinitiation chemistries is known by the tradename Irgacure. Irgacure and Darocur are tradenames for a family of photoinitiator chemistries that include alpha-Hydroxyketone, Phenylglyoxylate, Benzyldimethyl-ketal and alpha-Aminoketone. These photoinitiators have been used extensively in the industrial manufacturing of coatings, paints, epoxies and other construction material surface treatments. The major manufacturer and distributor of Irgacure and Darocur is the Ciba Specialty Chemicals Company, (Tarrytown, NY). The most commonly used Irgacure for biological applications is Irgacure 2959, with the chemical structure of 2-Hydroxy-1-[4-(2-hydroxyethoxy) phenyl]-2-methyl-1-propanone. Irgacure has a major energy absorption peak in the ultraviolet energy range with a maximum at 276 nm. Irgacure 2959 has been used previously in tissue engineering and now bioprinting applications as a technique to rapidly cross link solutions to form a gel [159, 322–324, 341–343].

One of the drawbacks of using Irgacure 2959, as well as many other Irgacures, is the need to use UV wavelength light (276 nm) to support rapid photoinitiation and crosslinking of materials. Ultraviolet light has been shown to have significant negative effects on cell function including cell death and mutagenesis [344–347]. It should be carefully noted that most of these earlier studies evaluating the toxic effect of UV light on cell function were performed in 2D cultures systems with cells growing as monolayers on the bottom of tissue culture plates. During UV light cross linking of materials during bioprinting the cells are exposed to UV light in 3D gels. UV sensitivity has been shown to be cell type specific [341]. Many other Irgacures have been evaluated in tissue engineering and bioprinting applications including; Irgacure 165, 184 and 907 [342, 348, 349]. The deleterious effect of Irgacures may be cell specific and polymer specific and thus, their use in bioprinting remains promising and the subject of additional study.

Other UV photoinitiators include; **2-isocyanatoethyl methacrylate** used to photocrosslink fish elastin [350], **benzoyl benzylamine** [351],

Alternative visible light photoinitiators are under study [186, 304, 325, 352]. Bahney and colleagues have done significant work evaluating novel visible light activated photoinitiators and cross linkers to avoid the previously noted deleterious effects of UV cross linkers [325]. As pointed out by these authors a major advantage of Irgacure 2959 is that it is a type I photoinitiator (cleavage type), and only a single molecules is necessary for both the photoactivation and initiation of the crosslinking reaction. As a contrast many of the visible light system (often utilizing type II photoinitiators [338]) require two or more reagents [78, 325, 348, 353]. Irgacure 2959 does have an absorbance peak at 365 nm, however, crosslinking at this wavelength requires extended exposure times.

Some of the visible light photoinitiators include **Camphorquinone** [348, 354], **Thiol-norbornene (thiol-ene)** [355], **eosin Y** [176, 184, 325, 356–360], **riboflavin** [361], **Lucirin-TPO** [362], **Rose Bengal/furfuryl** [363, 364], **Ethyl Eosin** [325, 357, 360, 365, 366], **methacrylic anhydride** [367], **2,2-dimethoxy-2-phenylacetophenone** [348]. These photoinitiators are most often used with other chemistries in Type II photoinitiation reactions (Table 2).

Table 2 Photoiniator Chemistries Used in Bioinks

Photoinitiator	Chemical structure	Wavelength (nm)
VA-086	2,2′-azobis[2-methyl-N-(2-hydroxyethyl) propionamide]	365
IRGACURE 2959	2-Hydroxy-1-[4-(2-hydroxyethoxy) phenyl]-2-methyl-1-propanone	276
DMAP	2,2-dimethoxy-2-phenylacetophenone	450
Rose Bengal	4,5,6,7-tetrachloro-2′,4′,5′,7′-tetraiodofluorescein	549
Eosin-Y	2-(2,4,5,7-tetrabromo-6-oxido-3-oxo-3*H*-xanthen-9-yl)benzoate	510
Ethyl Eosin	2′,4′,5′,7′-Tetrabromoeosin	532
Camphorquinone	Camphorquinone	467
Methacrylic anhydride	Methacrylic anhydride	530

1.7 Summary

The field of 3D Bioprinting is expanding rapidly with improvements in equipment, software for designing structures, delivery pens and systems. An integral part of bioprinting technology is the mixture of materials, cells and additives the field defines as Bioinks. These inks are based on simple naturally occurring substances but have now begun to evolve into new biomaterials with the ability to support tissue maturation following 3D printing. These new bioinks will support the development of new technologies that will rapidly move bioprinting into a commonly used technology in the laboratory and in the clinic.

References

1. Hutmacher DW, et al. Mechanical properties and cell cultural response of polycaprolactone scaffolds designed and fabricated via fused deposition modeling. J Biomed Mater Res. 2001;55:203–16.
2. Charest K, Mak-Jurkauskas ML, Cinicola D, Clausen AM. Fused deposition modeling provides solution for magnetic resonance imaging of solid dosage form by advancing design quickly from prototype to final product. J Lab Autom. 2013;18:63–8.
3. Beguma Z, Chhedat P. Rapid prototyping—when virtual meets reality. Int J Comput Dent. 2014;17:297–306.
4. Mironov V, Boland T, Trusk T, Forgacs G, Markwald RR. Organ printing: computer-aided jet-based 3D tissue engineering. Trends Biotechnol. 2003;21:157–61.
5. Siemens-Wapniarski WJ, Givens MP. The experimental production of synthetic holograms. Appl Opt. 1968;7:535–8.
6. Bartels KA, Bovik AC, Crawford RC, Diller KR, Aggarwal SJ. Selective laser sintering for the creation of solid models from 3D microscopic images. Biomed Sci Instrum. 1993;29:243–50.
7. Dyson JA, Genever PG, Dalgarno KW, Wood DJ. Development of custom-built bone scaffolds using mesenchymal stem cells and apatite-wollastonite glass-ceramics. Tissue Eng. 2007;13:2891–901.
8. Wanner M, et al. Effects of non-invasive, 1210 nm laser exposure on adipose tissue: results of a human pilot study. Lasers Surg Med. 2009;41:401–7.
9. Kim KB, Kim JH, Kim WC, Kim JH. Three-dimensional evaluation of gaps associated with fixed dental prostheses fabricated with new technologies. J Prosthet Dent. 2014;112:1432–6.
10. Lohfeld S, Cahill S, Doyle H, McHugh PE. Improving the finite element model accuracy of tissue engineering scaffolds produced by selective laser sintering. J Mater Sci Mater Med. 2015;26:5376.
11. Cvetkovic C, et al. Three-dimensionally printed biological machines powered by skeletal muscle. Human cartilage tissue fabrication using three-dimensional inkjet printing technology. Proc Natl Acad Sci U S A. 2014;111:10125–30.
12. Reyes A, Turkyilmaz I, Prihoda TJ. Accuracy of surgical guides made from conventional and a combination of digital scanning and rapid prototyping techniques. J Prosthet Dent. 2015;113(4):295–303.
13. Roth EA, et al. Inkjet printing for high-throughput cell patterning. Biomaterials. 2004;25:3707–15.
14. Cooper GM, et al. Inkjet-based biopatterning of bone morphogenetic protein-2 to spatially control calvarial bone formation. Tissue Eng Part A. 2010;16:1749–59.

15. Phillippi JA, et al. Microenvironments engineered by inkjet bioprinting spatially direct adult stem cells toward muscle- and bone-like subpopulations. Stem Cells. 2008;26:127–34.
16. Campbell PG, Weiss LE. Tissue engineering with the aid of inkjet printers. Expert Opin Biol Ther. 2007;7:1123–7.
17. Xu T, Jin J, Gregory C, Hickman JJ, Boland T. Inkjet printing of viable mammalian cells. Biomaterials. 2005;26:93–9.
18. Smith CM, Christian JJ, Warren WL, Williams SK. Characterizing environmental factors that impact the viability of tissue-engineered constructs fabricated by a direct-write bioassembly tool. Tissue Eng. 2007;13:373–83.
19. Smith CM, Cole Smith J, Williams SK, Rodriguez JJ, Hoying JB. Automatic thresholding of three-dimensional microvascular structures from confocal microscopy images. J Microsc. 2007;225:244–57.
20. Smith CM, et al. Three-dimensional bioassembly tool for generating viable tissue-engineered constructs. Tissue Eng. 2004;10:1566–76.
21. Chang CC, et al. Determinants of microvascular network topologies in implanted neovasculatures. Arterioscler Thromb Vasc Biol. 2012;32:5–14.
22. Chang CC, Boland ED, Williams SK, Hoying JB. Direct-write bioprinting three-dimensional biohybrid systems for future regenerative therapies. J Biomed Mater Res B Appl Biomater. 2011;98:160–70.
23. Chang R, Emami K, Wu H, Sun W. Biofabrication of a three-dimensional liver micro-organ as an in vitro drug metabolism model. Biofabrication. 2010;2:045004.
24. Chang R, Nam J, Sun W. Direct cell writing of 3D microorgan for in vitro pharmacokinetic model. Tissue Eng Part C Methods. 2008;14:157–66.
25. Mironov V, Kasyanov V, Markwald RR. Nanotechnology in vascular tissue engineering: from nanoscaffolding towards rapid vessel biofabrication. Trends Biotechnol. 2008;26:338–44.
26. Mironov V. Toward human organ printing: Charleston Bioprinting Symposium. ASAIO J. 2006;52:e27–30.
27. Mironov V, Drake C, Wen X. Research project: Charleston Bioengineered Kidney Project. Biotechnol J. 2006;1:903–5.
28. Jakab K, Damon B, Neagu A, Kachurin A, Forgacs G. Three-dimensional tissue constructs built by bioprinting. Biorheology. 2006;43:509–13.
29. Mironov V, Reis N, Derby B. Review: bioprinting: a beginning. Tissue Eng. 2006;12:631–4.
30. Catros S, et al. Layer-by-layer tissue microfabrication supports cell proliferation in vitro and in vivo. Tissue Eng Part C Methods. 2012;18:62–70.
31. Sides CR, et al. Probing the 3-D structure, dynamics, and stability of bacterial collagenase collagen binding domain (apo- versus holo-) by limited proteolysis MALDI-TOF MS. J Am Soc Mass Spectrom. 2012;23:505–19.
32. Gruene M, Unger C, Koch L, Deiwick A, Chichkov B. Dispensing pico to nanolitre of a natural hydrogel by laser-assisted bioprinting. Biomed Eng Online. 2011;10:19.
33. Guillotin B, Guillemot F. Cell patterning technologies for organotypic tissue fabrication. Trends Biotechnol. 2011.
34. Guillotin B, et al. Laser assisted bioprinting of engineered tissue with high cell density and microscale organization. Biomaterials. 2010;31:7250–6.
35. Tirella A, Ahluwalia A. The impact of fabrication parameters and substrate stiffness in direct writing of living constructs. Biotechnol Prog. 2012;28:1315–20.
36. Parzel CA, Pepper ME, Burg T, Groff RE, Burg KJ EDTA enhances high-throughput two-dimensional bioprinting by inhibiting salt scaling and cell aggregation at the nozzle surface. J Tissue Eng Regen Med. 2009;3:260–8.
37. Khalil S, Sun W. Bioprinting endothelial cells with alginate for 3D tissue constructs. J Biomech Eng. 2009;131:111002.
38. Chien KB, et al. Three-dimensional printing of soy protein scaffolds for tissue regenerationIn vivo acute and humoral response to three-dimensional porous soy protein scaffolds. Tissue Eng Part C Methods. 2013;19:417–26.

39. Zhang Y, Yu Y, Chen H, Ozbolat IT. Characterization of printable cellular micro-fluidic channels for tissue engineering. Biofabrication. 2013;5:025004.
40. Buyukhatipoglu K, Jo W, Sun W, Clyne AM. The role of printing parameters and scaffold biopolymer properties in the efficacy of a new hybrid nano-bioprinting system. Biofabrication. 2009;1:035003.
41. Lindbergh CA. A culture flask for the circulation of a large quantity of fluid medium. J Exp Med. 1939;70:231–8.
42. Carrel A, Lindbergh CA. The culture of whole organs. Science. 1935;81:621–3.
43. Lindbergh CA. An apparatus for the culture of whole organs. J Exp Med. 1935;62:409–31.
44. Lindbergh CA. A method for washing corpuscles in suspension. Science. 1932;75:415–6.
45. Madri JA, Williams SK. Capillary endothelial cell cultures: phenotypic modulation by matrix components. J Cell Biol. 1983;97:153–65.
46. Montesano R, Orci L, Vassalli P. In vitro rapid organization of endothelial cells into capillary-like networks is promoted by collagen matrices. J Cell Biol. 1983;97:1648–52.
47. Folkman J, Hochberg M. Self-regulation of growth in three dimensions. J Exp Med. 1973;138:745–53.
48. Elsdale T, Bard J. Collagen substrata for studies on cell behavior. J Cell Biol. 1972;54:626–37.
49. Cole KA, Krizman DB, Emmert-Buck MR. The genetics of cancer–a 3D model. NatGenet. 1999;21:38–41.
50. Haas TL, Davis SJ, Madri JA. Three-dimensional type I collagen lattices induce coordinate expression of matrix metalloproteinases MT1-MMP and MMP-2 in microvascular endothelial cells. J Biol Chem. 1998;273:3604–10.
51. Carter WB, Crowell SL, Boswell CA, Williams SK. Stimulation of angiogenesis by canine parathyroid tissue. Surgery. 1996;120:1089–94.
52. Hoying JB, Boswell CA, Williams SK. Angiogenic potential of microvessel fragments established in three-dimensional collagen gels. In Vitro Cell Dev Biol Anim. 1996;32:409–19.
53. Perrone CE, Fenwick-Smith D, Vandenburgh HH. Collagen and stretch modulate autocrine secretion of insulin-like growth factor-1 and insulin-like growth factor binding proteins from differentiated skeletal muscle cells. J Biol Chem. 1995;270:2099–106.
54. Roguet R, Regnier M, Cohen C, Dossou KG, Rougier A. The use of in vitro reconstituted human skin in dermotoxicity testing. Toxicol In Vitro. 1994;8:635–9.
55. Hikichi Y, Sugihara H, Sugimoto E. Differentiation of brown adipose cells in three-dimensional collagen gel culture. Pathol Res Pract. 1993;189:73–82.
56. Nicosia RF, Bonanno E, Smith M. Fibronectin promotes the elongation of microvessels during angiogenesis in vitro. J Cell Physiol. 1993;154:654–61.
57. Villaschi S, Nicosia RF. Angiogenic role of endogenous basic fibroblast growth factor released by rat aorta after injury. Am J Pathol. 1993;143:181–90.
58. Madri JA, Pratt BM, Tucker AM. Phenotypic modulation of endothelial cells by transforming growth factor-beta depends upon the composition and organization of the extracellular matrix. J Cell Biol. 1988;106:1375–84.
59. Emerman JT, Burwen SJ, Pitelka DR. Substrate properties influencing ultrastructural differentiation of mammary epithelial cells in culture. Tissue Cell. 1979;11:109–19.
60. Park A, Wu B, Griffith LG. Integration of surface modification and 3D fabrication techniques to prepare patterned poly(L-lactide) substrates allowing regionally selective cell adhesion. J Biomater Sci Polym Ed. 1998;9:89–110.
61. Kim K, Lee CH, Kim BK, Mao JJ. Anatomically shaped tooth and periodontal regeneration by cell homing. J Dent Res. 2010;89:842–7.
62. Nakamura M, Iwanaga S, Henmi C, Arai K, Nishiyama Y. Biomatrices and biomaterials for future developments of bioprinting and biofabrication. Biofabrication. 2010;2:014110.
63. Wilson WC Jr, Boland T. Cell and organ printing 1: protein and cell printers. Anat Rec A Discov Mol Cell Evol Biol. 2003;272:491–6.
64. Rees DA. Structure, conformation, and mechanism in the formation of polysaccharide gels and networks. Adv Carbohydr Chem Biochem. 1969;24:267–332.

65. Young AA, Legrice IJ, Young MA, Smaill BH. Extended confocal microscopy of myocardial laminae and collagen network. J Microsc. 1998;192(Pt 2):139–50.
66. Beavis MJ, et al. Human peritoneal fibroblast proliferation in 3-dimensional culture: modulation by cytokines, growth factors and peritoneal dialysis effluent. Hydrogel-based three-dimensional matrix for neural cells. Kidney Int. 1997;51:205–15.
67. Bellamkonda R, Ranieri JP, Bouche N, Aebischer P. Hydrogel-based three-dimensional matrix for neural cells. J Biomed Mater Res. 1995;29:663–71.
68. Bittinger F, et al. Reconstruction of peritoneal-like structure in three-dimensional collagen gel matrix culture. Human peritoneal fibroblast proliferation in 3-dimensional culture: modulation by cytokines, growth factors and peritoneal dialysis effluent. Hydrogel-based three-dimensional matrix for neural cells. Exp Cell Res. 1997;236:155–60.
69. Blaeser A, et al. Biofabrication under fluorocarbon: a novel freeform fabrication technique to generate high aspect ratio tissue-engineered constructs. Biores Open Access. 2013;2:374–84.
70. Baldwin D, Crane V, Rice D. A comparison of gel-based, nylon filter and microarray techniques to detect differential RNA expression in plants. Curr Opin Plant Biol. 1999;2:96–103.
71. Norotte C, Marga FS, Niklason LE, Forgacs G. Scaffold-free vascular tissue engineering using bioprinting. Biomaterials. 2009;30:5910–7.
72. Duarte Campos DF, et al. Three-dimensional printing of stem cell-laden hydrogels submerged in a hydrophobic high-density fluid. Biofabrication. 2013;5:015003.
73. Stokes CL, Rupnick MA, Williams SK, Lauffenburger DA. Chemotaxis of human microvessel endothelial cells in response to acidic fibroblast growth factor. Lab Invest. 1990;63:657–68.
74. Hoying JB, Williams SK. Measurement of endothelial cell migration using an improved linear migration assay. Presented at the 1995 Microcirculatory Society Meeting. Microcirculation. 1996;3:167–74.
75. Rupnick MA, Stokes CL, Williams SK, Lauffenburger DA. Quantitative analysis of random motility of human microvessel endothelial cells using a linear under-agarose assay. Lab Invest. 1988;59:363–72.
76. Deryckere D, Eeckhaut T, Van Huylenbroeck J, Van Bockstaele E. Low melting point agarose beads as a standard method for plantlet regeneration from protoplasts within the Cichorium genus. Plant Cell Rep. 2012;31:2261–9.
77. Desroches BR, et al. Functional scaffold-free 3-D cardiac microtissues: a novel model for the investigation of heart cells. Am J Physiol Heart Circ Physiol. 2012;302:H2031–42.
78. Skardal A, Zhang J, Prestwich GD. Bioprinting vessel-like constructs using hyaluronan hydrogels crosslinked with tetrahedral polyethylene glycol tetracrylates. Biomaterials. 2010;31:6173–81.
79. Lee VK, et al. Generation of multiscale vascular network system within 3D hydrogel using 3D bioprinting technology. Cell Mol Bioeng. 2014;7:460–72.
80. Visconti RP, et al. Towards organ printing: engineering an intra-organ branched vascular tree. Expert Opin Biol Ther. 2010;10:409–20.
81. Chong JA, Chong MP, Docking AR. The surface of gypsum cast in alginate impressions. Dent Pract Dent Rec. 1965;16:107–9.
82. Moshaverinia A, et al. Alginate hydrogel as a promising scaffold for dental-derived stem cells: an in vitro study. J Mater Sci Mater Med. 2012;23:3041–51.
83. Smith CA. A new and effective hemostatic agent. Science. 1946;103:634.
84. Blaine G. experimental observations on absorbable alginate products in surgery: gel, film, gauze and foam. Ann Surg. 1947;125:102–14.
85. Mallory GD Jr. Alginate impression technique. U S Nav Med Bull. 1946;46:1083–7.
86. Hunter SK, Kao JM, Wang Y, Benda JA, Rodgers VG. Promotion of neovascularization around hollow fiber bioartificial organs using biologically active substances. ASAIO J. 1999;45:37–40.
87. Hwang CM, et al. Fabrication of three-dimensional porous cell-laden hydrogel for tissue engineering. Biofabrication. 2010;2:035003.
88. Duan B, Hockaday LA, Kang KH, Butcher JT. 3D bioprinting of heterogeneous aortic valve conduits with alginate/gelatin hydrogels. J Biomed Mater Res A. 2013.

89. Huang X, Zhang X, Wang X, Wang C, Tang B. Microenvironment of alginate-based microcapsules for cell culture and tissue engineering. J Biosci Bioeng. 2012;114:1–8.
90. Lee BR, et al. In situ formation and collagen-alginate composite encapsulation of pancreatic islet spheroids. Biomaterials. 2012;33:837–45.
91. Hals IK, Rokstad AM, Strand BL, Oberholzer J, Grill V. Alginate microencapsulation of human islets does not increase susceptibility to acute hypoxia. J Diabetes Res. 2013;2013:374925.
92. Gasperini L, Mano JF, Reis RL. Natural polymers for the microencapsulation of cells. J R Soc Interface. 2014;11:20140817.
93. Jitraruch S, et al. Alginate microencapsulated hepatocytes optimised for transplantation in acute liver failure Alginate encapsulation of human embryonic stem cells to enhance directed differentiation to pancreatic islet-like cells. PLoS One. 2014;9:e113609.
94. Kryukov O, et al. Three-dimensional perfusion cultivation of human cardiac-derived progenitors facilitates their expansion while maintaining progenitor state. Alginate microencapsulation of human islets does not increase susceptibility to acute hypoxia. Tissue Eng Part C Methods. 2014;20:886–94.
95. Williams SK, Touroo JS, Church KH, Hoying JB. Encapsulation of adipose stromal vascular fraction cells in alginate hydrogel spheroids using a direct-write three-dimensional printing system. Biores Open Access. 2013;2:448–54.
96. Matyash M, et al. Novel soft alginate hydrogel strongly supports neurite growth and protects neurons against oxidative stress. Tissue Eng Part A. 2012;18:55–66.
97. Heidari R, et al. Alginate as a cell culture substrate for growth and differentiation of human retinal pigment epithelial cells. Appl Biochem Biotechnol. 2014;175(5):2399–412.
98. Buyukhatipoglu K, Chang R, Sun W, Clyne AM. Bioprinted nanoparticles for tissue engineering applications. Tissue Eng Part C Methods. 2010;16:631–42.
99. Ueno M, Oda T. Biological activities of alginate. Adv Food Nutr Res. 2014;72:95–112.
100. Selimoglu SM, Ayyildiz-Tamis D, Gurhan ID, Elibol M. Purification of alginate and feasible production of monoclonal antibodies by the alginate-immobilized hybridoma cells. J Biosci Bioeng. 2012;113:233–8.
101. Badur AH, et al. Characterization of the alginate lyases from Vibrio splendidus 12B01. Appl Environ Microbiol. 2015.
102. Diaz-Barrera A, Martinez F, Pezoa FG, Acevedo F. Evaluation of gene expression and alginate production in response to oxygen transfer in continuous culture of Azotobacter vinelandii. PLoS One. 2014;9:e105993.
103. Morris ER, Rees DA. Principles of biopolymer gelation. Possible models for mucus gel structure. Br Med Bull. 1978;34:49–53.
104. Morris ER, Rees DA. & Robinson, G. Cation-specific aggregation of carrageenan helices: Domain model of polymer gel structure. J Mol Biol. 1980;138:349–62.
105. Blandino A, Macias M, Cantero D. Formation of calcium alginate gel capsules: influence of sodium alginate and $CaCl_2$ concentration on gelation kinetics. J Biosci Bioeng. 1999;88:686–9.
106. Siew CK, Williams PA, Young NW. New insights into the mechanism of gelation of alginate and pectin: charge annihilation and reversal mechanism. Biomacromolecules. 2005;6:963–9.
107. Madihally SV, Matthew HW. Porous chitosan scaffolds for tissue engineering [In Process Citation]. Biomaterials. 1999;20:1133–42.
108. Chupa JM, Foster AM, Sumner SR, Madihally SV, Matthew HW. Vascular cell responses to polysaccharide materials: in vitro and in vivo evaluations. Biomaterials. 2000;21:2315–22.
109. Risbud MV, Bhonde RR. Islet immunoisolation: experience with biopolymers. J Biomater Sci Polym Ed. 2001;12:1243–52.
110. Wang YC, Ho CC. Micropatterning of proteins and mammalian cells on biomaterials. FASEB J. 2004;18:525–7.
111. Co CC, Wang YC, Ho CC. Biocompatible micropatterning of two different cell types. J Am Chem Soc. 2005;127:1598–9.

112. Patel M, et al. Video-gait analysis of functional recovery of nerve repaired with chitosan nerve guides. Tissue Eng. 2006;12:3189–99.
113. Ficke JR, Pollak AN. Extremity war injuries: development of clinical treatment principles. J Am Acad Orthop Surg. 2007;15:590–5.
114. Chen B, et al. Dynamics of smooth muscle cell deadhesion from thermosensitive hydroxybutyl chitosan. Biomaterials. 2007;28:1503–14.
115. Fernandez JG, et al. Micro- and nanostructuring of freestanding, biodegradable, thin sheets of chitosan via soft lithography. J Biomed Mater Res A. 2008;85:242–7.
116. Morgan SM, et al. Formation of a human-derived fat tissue layer in P(DL)LGA hollow fibre scaffolds for adipocyte tissue engineering. Biomaterials. 2009;30:1910–7.
117. Altman AM, et al. IFATS collection: Human adipose-derived stem cells seeded on a silk fibroin-chitosan scaffold enhance wound repair in a murine soft tissue injury model. Stem Cells. 2009;27:250–8.
118. Ju C, et al. Antidiabetic effect and mechanism of chitooligosaccharides. Biol Pharm Bull. 2010;33:1511–6.
119. Xu HH, Zhao L, Weir MD. Stem cell-calcium phosphate constructs for bone engineering. J Dent Res. 2010;89:1482–8.
120. Zhao L, Weir MD, Xu HH. Human umbilical cord stem cell encapsulation in calcium phosphate scaffolds for bone engineering. Biomaterials. 2010;31:3848–57.
121. Reddy AM, et al. In vivo tracking of mesenchymal stem cells labeled with a novel chitosan-coated superparamagnetic iron oxide nanoparticles using 3.0T MRI. J Korean Med Sci. 2010;25:211–9.
122. Huang GS, Dai LG, Yen BL, Hsu SH. Spheroid formation of mesenchymal stem cells on chitosan and chitosan-hyaluronan membranes. Biomaterials. 2011;32:6929–45.
123. Natesan S, et al. A bilayer construct controls adipose-derived stem cell differentiation into endothelial cells and pericytes without growth factor stimulation. Tissue Eng Part A. 2011;17:941–53.
124. Grigolo B, et al. Chemical-physical properties and in vitro cell culturing of a novel biphasic biomimetic scaffold for osteochondral tissue regeneration. J Biol Regul Homeost Agents. 2011;25:3–13.
125. Liao HT, Chen CT, Chen JP. Osteogenic differentiation and ectopic bone formation of canine bone marrow-derived mesenchymal stem cells in injectable thermo-responsive polymer hydrogel. Tissue Eng Part C Methods. 2011;17:1139–49.
126. Lee JH, et al. Effect of fulminant hepatic failure porcine plasma supplemented with essential components on encapsulated rat hepatocyte spheroids. Transplant Proc. 2012;44:1009–11.
127. Qi BW, Yu AX, Zhu SB, Zhou M, Wu G. Chitosan/poly(vinyl alcohol) hydrogel combined with Ad-hTGF-beta1 transfected mesenchymal stem cells to repair rabbit articular cartilage defects. Exp Biol Med (Maywood). 2013;238:23–30.
128. Phan-Lai V, et al. Three-dimensional scaffolds to evaluate tumor associated fibroblast-mediated suppression of breast tumor specific T cells. Biomacromolecules. 2013.
129. Huang F, et al. Preparation of three-dimensional macroporous chitosan-gelatin B microspheres and HepG2-cell culture. J Tissue Eng Regen Med. 2014.
130. de Vos P, Lazarjani HA, Poncelet D, Faas MM. Polymers in cell encapsulation from an enveloped cell perspective. Adv Drug Deliv Rev. 2014;67-68:15–34.
131. Duarte Campos DF, et al. The stiffness and structure of three-dimensional printed hydrogels direct the differentiation of mesenchymal stromal cells toward adipogenic and osteogenic lineages. Tissue Eng Part A. 2014.
132. Belscak-Cvitanovic A, et al. Improving the controlled delivery formulations of caffeine in alginate hydrogel beads combined with pectin, carrageenan, chitosan and psyllium. Food Chem. 2015;167:378–86.
133. Martin MJ, et al. Development of alginate microspheres as nystatin carriers for oral mucosa drug delivery. Carbohydr Polym. 2015;117:140–9.

134. Geng L, et al. Direct writing of chitosan scaffolds using a robotic system. Rapid Prototyp J. 2005;11:90–7.
135. Amsden BG, Sukarto A, Knight DK, Shapka SN. Methacrylated glycol chitosan as a photopolymerizable biomaterial. Biomacromolecules. 2007;8:3758–66.
136. Eccles SA, Purvies HP, Barnett SC, Alexander P. Inhibition of growth and metastasis of syngeneic transplantable tumours by an aromatic retinoic acid analogue. 2. T cell dependence of retinoid effects in vivo. Cancer Immunol Immunother. 1985;19:115–20.
137. Ahmed AB, Adel M, Karimi P, Peidayesh M. Pharmaceutical, cosmeceutical, and traditional applications of marine carbohydrates. Adv Food Nutr Res. 2014;73:197–220.
138. Sudha PN, Aisverya S, Nithya R, Vijayalakshmi K. Industrial applications of marine carbohydrates. Adv Food Nutr Res. 2014;73:145–81.
139. Mihaila SM, Popa EG, Reis RL, Marques AP, Gomes ME. Fabrication of endothelial cell-laden carrageenan microfibers for microvascularized bone tissue engineering applications. Biomacromolecules. 2014;15:2849–60.
140. Tsubaki S, et al. Effects of acidic functional groups on dielectric properties of sodium alginates and carrageenans in water. Carbohydr Polym. 2015;115:78–87.
141. Sharma A, Bhat S, Vishnoi T, Nayak V, Kumar A. Three-dimensional supermacroporous carrageenan-gelatin cryogel matrix for tissue engineering applications. Biomed Res Int. 2013;2013:478279.
142. Ehrmann RL, Gey GO. The growth of cells on a transparent gel of reconstituted rat-tail collagen. J Natl Cancer Inst. 1956;16:1375–403.
143. Yang J, et al. Primary culture of human mammary epithelial cells embedded in collagen gels. J Natl Cancer Inst. 1980;65:337–43.
144. Talley DJ, Roy WA, Li JJ. Behavior of primary and serially transplanted estrogen-dependent renal carcinoma cells in monolayer and in collagen gel culture. In Vitro. 1982;18:149–56.
145. Lawler EM, Miller FR, Heppner GH. Significance of three-dimensional growth patterns of mammary tissues in collagen gels. In Vitro. 1983;19:600–10.
146. Chang CC, Hoying JB. Directed three-dimensional growth of microvascular cells and isolated microvessel fragments. Cell Transplant. 2006;15:533–40.
147. Rauterberg J, Kuhn K. Acid soluble calf skin collagen. Characterization of the peptides obtained by cyanogen bromide cleavage of its alpha-1-chain. Eur J Biochem. 1971;19:398–407.
148. Chung E, Rhodes K, Miller EJ. Isolation of three collagenous components of probable basement membrane origin from several tissues. Biochem Biophys Res Commun. 1976;71:1167–74.
149. Williams SK, et al. Adult human endothelial cell compatibility with prosthetic graft material. J Surg Res. 1985;38:618–29.
150. Baker KS, Williams SK, Jarrell BE, Koolpe EA, Levine E. Endothelialization of human collagen surfaces with human adult endothelial cells. Am J Surg. 1985;150:197–200.
151. Gallop PM, Seifter S, Meilman E. Studies on collagen. I. The partial purification, assay, and mode of activation of bacterial collagenase. J Biol Chem. 1957;227:891–906.
152. Chandrakasan G, Torchia DA, Piez KA. Preparation of intact monomeric collagen from rat tail tendon and skin and the structure of the nonhelical ends in solution. J Biol Chem. 1976;251:6062–7.
153. Lee W, et al. Three-dimensional bioprinting of rat embryonic neural cells. Neuroreport. 2009;20:798–803.
154. Moon S, et al. Layer by layer three-dimensional tissue epitaxy by cell-laden hydrogel droplets. Tissue Eng Part C Methods. 2010;16:157–66.
155. Chang CC., et al. In vitro patterned microvessels lose alignment in vivo. in Microcirculatory Society Meeting. 2010.
156. Pepper ME, et al. Post-bioprinting processing methods to improve cell viability and pattern fidelity in heterogeneous tissue test systems. Conf Proc IEEE Eng Med Biol Soc. 2010;2010:259–62.
157. Hong S, et al. Cellular behavior in micropatterned hydrogels by bioprinting system depended on the cell types and cellular interaction. J Biosci Bioeng. 2013;116:224–30.

158. Gao G, et al. Bioactive nanoparticles stimulate bone tissue formation in bioprinted three-dimensional scaffold and human mesenchymal stem cells. Three-dimensionally printed biological machines powered by skeletal muscle. Human cartilage tissue fabrication using three-dimensional inkjet printing technology. Biotechnol J. 2014;9:1304–11.
159. Cui X, Gao G, Yonezawa T, Dai G. Human cartilage tissue fabrication using three-dimensional inkjet printing technology. J Vis Exp. 2014.
160. Park JY, et al. A comparative study on collagen type I and hyaluronic acid dependent cell behavior for osteochondral tissue bioprinting. Biofabrication. 2014;6:035004.
161. Lee V, et al. Design and fabrication of human skin by three-dimensional bioprinting. Tissue Eng Part C Methods. 2014;20:473–84.
162. Li Y, Douglas EP. Effects of various salts on structural polymorphism of reconstituted type I collagen fibrils. Colloids Surf B Biointerfaces. 2013;112:42–50.
163. Li J, Zhang YP, Kirsner RS. Angiogenesis in wound repair: angiogenic growth factors and the extracellular matrix. Microsc Res Tech. 2003;60:107–114.
164. Kidd KR, Williams SK. Laminin-5-enriched extracellular matrix accelerates angiogenesis and neovascularization in association with ePTFE. J Biomed Mater Res A. 2004;69:294–304.
165. Kidd KR, et al. Stimulated endothelial cell adhesion and angiogenesis with laminin-5 modification of expanded polytetrafluoroethylene. Tissue Eng. 2005;11:1379–91.
166. Crabtree B, Subramanian V. Behavior of endothelial cells on Matrigel and development of a method for a rapid and reproducible in vitro angiogenesis assay. In Vitro Cell Dev Biol Anim. 2007;43:87–94.
167. Folkman J, Haudenschild CC, Zetter BR. Long-term culture of capillary endothelial cells. Proc Natl Acad Sci U S A. 1979;76:5217–21.
168. Jarrell B, et al. Human adult endothelial cell growth in culture. J Vasc Surg. 1984;1:757–64.
169. Ingber DE, Madri JA, Folkman J. Endothelial growth factors and extracellular matrix regulate DNA synthesis through modulation of cell and nuclear expansion. In Vitro Cell DevBiol. 1987;23:387–94.
170. Pratt KJ, et al. Kinetics of endothelial cell-surface attachment forces. J Vasc Surg. 1988;7:591–9.
171. Kim NS, Kim SJ. Isolation and cultivation of microvascular endothelial cells from rat lungs: effects of gelatin substratum and serum. Yonsei Med J. 1991;32:303–14.
172. Itoh T, et al. Reduced angiogenesis and tumor progression in gelatinase A-deficient mice. Cancer Res. 1998;58:1048–51.
173. Ratnikov BI, Deryugina EI, Strongin AY. Gelatin zymography and substrate cleavage assays of matrix metalloproteinase-2 in breast carcinoma cells overexpressing membrane type-1 matrix metalloproteinase. Lab Invest. 2002;82:1583–90.
174. Varani J, et al. Vascular tube formation on matrix metalloproteinase-1-damaged collagen. Br J Cancer. 2008;98:1646–52.
175. Schiele NR, Chrisey DB, Corr DT. Gelatin-based laser direct-write technique for the precise spatial patterning of cells. Tissue Eng Part C Methods. 2011;17:289–98.
176. Qu T, Liu X. Nano-structured gelatin/bioactive glass hybrid scaffolds for the enhancement of odontogenic differentiation of human dental pulp stem cells. J Mater Chem B Mater Biol Med. 2013;1:4764–72.
177. Occhetta P, et al. Fabrication of 3D cell-laden hydrogel microstructures through photomold patterning. Biofabrication. 2013;5:035002.
178. Brochhausen C, et al. Phenotypic redifferentiation and cell cluster formation of cultured human articular chondrocytes in a three-dimensional oriented gelatin scaffold in the presence of PGE2–first results of a pilot study. J Biomed Mater Res A. 2013;101:2374–82.
179. Das S, et al. Enhanced redifferentiation of chondrocytes on microperiodic silk/gelatin scaffolds: toward tailor-made tissue engineering. Biomacromolecules. 2013;14:311–21.
180. Garcia Cruz DM, Sardinha V, Escobar Ivirico JL, Mano JF, Gomez Ribelles JL. Gelatin microparticles aggregates as three-dimensional scaffolding system in cartilage engineering. J Mater Sci Mater Med. 2013;24:503–13.

181. Focaroli S, et al. Chondrogenic differentiation of human adipose mesenchimal stem cells: influence of a biomimetic gelatin genipin crosslinked porous scaffold. Microsc Res Tech. 2014;77:928–34.
182. Fang JY, Tan SJ, Yang Z, Tayag C, Han B. Tumor bioengineering using a transglutaminase crosslinked hydrogel. PLoS One. 2014;9:e105616.
183. Levato R, et al. Biofabrication of tissue constructs by 3D bioprinting of cell-laden microcarriers. Biofabrication. 2014;6:035020.
184. Quan R, Zheng X, Xu S, Zhang L, Yang D. Gelatin-chondroitin-6-sulfate-hyaluronic acid scaffold seeded with vascular endothelial growth factor 165 modified hair follicle stem cells as a three-dimensional skin substitute. Stem Cell Res Ther. 2014;5:118.
185. Sachar A, et al. Cell-matrix and cell-cell interactions of human gingival fibroblasts on three-dimensional nanofibrous gelatin scaffolds. J Tissue Eng Regen Med. 2014;8:862–73.
186. Skardal A, et al. Photocrosslinkable hyaluronan-gelatin hydrogels for two-step bioprinting. Tissue Eng Part A. 2010;16:2675–85.
187. Halbleib M, Skurk T, de Luca C, von Heimburg D, Hauner H. Tissue engineering of white adipose tissue using hyaluronic acid-based scaffolds. I: in vitro differentiation of human adipocyte precursor cells on scaffolds. Biomaterials. 2003;24:3125–32.
188. Baumann L, Kaufman J, Saghari S. Collagen fillers. Dermatol Ther. 2006;19:134–40.
189. Sadick N, Sorhaindo L. The utility of soft tissue fillers in clinical dermatology: treatment of fine wrinkles and skin defects. Expert Rev Med Devices. 2007;4:559–65.
190. Mehlhorn AT, et al. Differential effects of BMP-2 and TGF-beta1 on chondrogenic differentiation of adipose derived stem cells. Cell Prolif. 2007;40:809–23.
191. Quatela VC, Chow J. Synthetic facial implants. Facial Plast Surg Clin North Am. 2008;16:1–10,v.
192. Lee H, Park TG. Photo-crosslinkable, biomimetic, and thermo-sensitive pluronic grafted hyaluronic acid copolymers for injectable delivery of chondrocytes. J Biomed Mater Res A. 2009;88:797–806.
193. Song SJ, et al. A three-dimensional bioprinting system for use with a hydrogel-based biomaterial and printing parameter characterization. Artif Organs. 2010;34:1044–8.
194. Wu SC, Chang JK, Wang CK, Wang GJ, Ho ML. Enhancement of chondrogenesis of human adipose derived stem cells in a hyaluronan-enriched microenvironment. Biomaterials. 2010;31:631–40.
195. Park H, et al. Three-dimensional hydrogel model using adipose-derived stem cells for vocal fold augmentation. Tissue Eng Part A. 2010;16:535–43.
196. Yoon IS, et al. Proliferation and chondrogenic differentiation of human adipose-derived mesenchymal stem cells in porous hyaluronic acid scaffold. J Biosci Bioeng. 2011.
197. Filardo G, et al. Platelet-rich plasma vs hyaluronic acid to treat knee degenerative pathology: study design and preliminary results of a randomized controlled trial. BMC Musculoskelet Disord. 2012;13:229.
198. Cerza F, et al. Comparison between hyaluronic acid and platelet-rich plasma, intraarticular infiltration in the treatment of gonarthrosis. Am J Sports Med. 2012;40:2822–7.
199. Murphy SV, Skardal A, Atala A. Evaluation of hydrogels for bioprinting applications. J Biomed Mater Res A. 2013;101:272–84.
200. Pradhan-Bhatt S, et al. Implantable three-dimensional salivary spheroid assemblies demonstrate fluid and protein secretory responses to neurotransmitters. Tissue Eng Part A. 2013;19:1610–20.
201. de Belder AN, Wik KO. Preparation and properties of fluorescein-labelled hyaluronate. Carbohydr Res. 1975;44:251–7.
202. Bashkin P, et al. Basic fibroblast growth factor binds to subendothelial extracellular matrix and is released by heparitinase and heparin-like molecules. Biochemistry. 1989;28:1737–43.
203. Hsu S, Jamieson AM, Blackwell J. Viscoelastic studies of extracellular matrix interactions in a model native collagen gel system. Biorheology. 1994;31:21–36.
204. Kito H, Matsuda T. Biocompatible coatings for luminal and outer surfaces of small-caliber artificial grafts. J Biomed Mater Res. 1996;30:321–30.

205. Pall EA, Bolton KM, Ervasti JM. Differential heparin inhibition of skeletal muscle alpha-dystroglycan binding to laminins. J Biol Chem. 1996;271:3817–21.
206. Rao SS, et al. Glioblastoma behaviors in three-dimensional collagen-hyaluronan composite hydrogels. ACS Appl Mater Interfaces. 2013;5:9276–84.
207. Choi JH, Jun JH, Kim JH, Sung HJ, Lee JH. Synergistic effect of interleukin-6 and hyaluronic acid on cell migration and ERK activation in human keratinocytes. J Korean Med Sci. 2014;(Suppl 3):210–6.
208. Zopf DA, et al. Computer aided-designed, 3-dimensionally printed porous tissue bioscaffolds for craniofacial soft tissue reconstruction. Otolaryngol Head Neck Surg. 2015;152:57–62.
209. Duan B, Kapetanovic E, Hockaday LA, Butcher JT. Three-dimensional printed trileaflet valve conduits using biological hydrogels and human valve interstitial cells. Acta Biomater. 2014;10:1836–46.
210. Bhardwaj N, Chakraborty S, Kundu SC. Freeze-gelled silk fibroin protein scaffolds for potential applications in soft tissue engineering. Int J Biol Macromol. 2011;49:260–7.
211. Shen W, et al. Allogenous tendon stem/progenitor cells in silk scaffold for functional shoulder repair. Cell Transplant. 2012;21:943–58.
212. Diab T, et al. A silk hydrogel-based delivery system of bone morphogenetic protein for the treatment of large bone defects. J Mech Behav Biomed Mater. 2012;11:123–31.
213. Kasoju N, Bora U. Silk fibroin in tissue engineering. Adv Healthc Mater. 2012;1:393–412.
214. Kundu B, Kundu SC. Bio-inspired fabrication of fibroin cryogels from the muga silkworm Antheraea assamensis for liver tissue engineering. Biomed Mater. 2013;8:055003.
215. Stoppel WL, Ghezzi CE, McNamara SL, Iii LD, Kaplan DL. Clinical applications of naturally derived biopolymer-based scaffolds for regenerative medicine. Ann Biomed Eng. 2014.
216. Das S, et al. Bioprintable, cell-laden silk fibroin-gelatin hydrogel supporting multilineage differentiation of stem cells for fabrication of three-dimensional tissue constructs. Acta Biomater. 2015;11:233–46.
217. Thitiwuthikiat P, Ii M, Saito T, Kanokpanont S, Tabata Y. A vascular patch prepared from Thai silk fibroin and gelatin hydrogel incorporating simvastatin-micelles to recruit endothelial progenitor cells. Tissue Eng Part A. 2014.
218. Bhaarathy V, et al. Biologically improved nanofibrous scaffolds for cardiac tissue engineering. Mater Sci Eng C Mater Biol Appl. 2014;44:268–77.
219. Panda N, Bissoyi A, Pramanik K, Biswas A. Development of novel electrospun nanofibrous scaffold from P. ricini and A. mylitta silk fibroin blend with improved surface and biological properties. Mater Sci Eng C Mater Biol Appl. 2015;48:521–32.
220. Bhardwaj N, et al. Correction: Silk fibroin-keratin based 3D scaffolds as a dermal substitute for skin tissue engineering. Integr Biol (Camb). 2015;7:142.
221. Zhang W. et al. Fabrication and in vivo implantation of ligament-bone composite scaffolds based on three-dimensional printing technique. Zhongguo Xiu Fu Chong Jian Wai Ke Za Zhi. 2014;28:314–7.
222. Schacht K, et al. Biofabrication of cell-loaded 3d spider silk constructs. Angew Chem Int Ed Engl.
223. Gamble JR, et al. Regulation of in vitro capillary tube formation by anti-integrin antibodies. J Cell Biol. 1993;121:931–43.
224. Vailhe B, Ronot X, Tracqui P, Usson Y, Tranqui L. In vitro angiogenesis is modulated by the mechanical properties of fibrin gels and is related to alpha(v)beta3 integrin localization. In Vitro Cell Dev Biol Anim. 1997;33:763–73.
225. Babaei S, et al. Role of nitric oxide in the angiogenic response in vitro to basic fibroblast growth factor. Circ Res. 1998;82:1007–15.
226. Koblizek TI, Weiss C, Yancopoulos GD, Deutsch U, Risau W. Angiopoietin-1 induces sprouting angiogenesis in vitro. Curr Biol. 1998;8:529–32.
227. Vailhe B, Lecomte M, Wiernsperger N, Tranqui L. The formation of tubular structures by endothelial cells is under the control of fibrinolysis and mechanical factors. Angiogenesis. 1998;2:331–44.

228. Korff T, Augustin HG Tensional forces in fibrillar extracellular matrices control directional capillary sprouting. J Cell Sci. 1999;112(Pt 19):3249–58.
229. Ribatti D, et al. Alterations of blood vessel development by endothelial cells overexpressing fibroblast growth factor-2. J Pathol. 1999;189:590–9.
230. Nakatsu MN, et al. Angiogenic sprouting and capillary lumen formation modeled by human umbilical vein endothelial cells (HUVEC) in fibrin gels: the role of fibroblasts and Angiopoietin-1. Microvasc Res. 2003;66:102–12.
231. Dietrich F, Lelkes PI. Fine-tuning of a three-dimensional microcarrier-based angiogenesis assay for the analysis of endothelial-mesenchymal cell co-cultures in fibrin and collagen gels. Angiogenesis. 2006;9:111–25.
232. Stephanou A, Meskaoui G, Vailhe B, Tracqui P. The rigidity in fibrin gels as a contributing factor to the dynamics of in vitro vascular cord formation. Microvasc Res. 2007;73:182–90.
233. Chen X, et al. Prevascularization of a fibrin-based tissue construct accelerates the formation of functional anastomosis with host vasculature. Tissue Eng Part A. 2009;15:1363–71.
234. Kniazeva E, Putnam AJ Endothelial cell traction and ECM density influence both capillary morphogenesis and maintenance in 3-D. Am J Physiol Cell Physiol. 2009;297:C179–87.
235. Carrion B, et al. Recreating the perivascular niche ex vivo using a microfluidic approach. Biotechnol Bioeng. 2010;107:1020–8.
236. Long JL, Zuk P, Berke GS, Chhetri DK. Epithelial differentiation of adipose-derived stem cells for laryngeal tissue engineering. Laryngoscope. 2010;120:125–31.
237. Zhao Z, Hao C, Zhao H, Liu J, Shao L. Injectable allogeneic bone mesenchymal stem cells: a potential minimally invasive therapy for atrophic nonunion. Med Hypotheses. 2011;77:912–3.
238. Muller M, Becher J, Schnabelrauch M, Zenobi-Wong M. Printing thermoresponsive reverse molds for the creation of patterned two-component hydrogels for 3D cell culture. J Vis Exp. 2013;e50632.
239. Jitraruch S, et al. Alginate microencapsulated hepatocytes optimised for transplantation in acute liver failure. Influence of modified alginate hydrogels on mesenchymal stem cells and olfactory bulb-derived glial cells cultures. PLoS One. 2014;9:e113609.
240. Palchesko RN, Szymanski JM, Sahu A, Feinberg AW. Shrink wrapping cells in a defined extracellular matrix to modulate the chemo-mechanical microenvironment. Cell Mol Bioeng. 2014;7:355–68.
241. Aplin AC, Nicosia RF, Nicosia RF, Tchao R, Leighton J. The rat aortic ring model of angiogenesis. Angiogenesis-dependent tumor spread in reinforced fibrin clot culture. Methods Mol Biol. 2015;1214:255–64.
242. Nishiyama Y, et al. Development of a three-dimensional bioprinter: construction of cell supporting structures using hydrogel and state-of-the-art inkjet technology. J Biomech Eng. 2009;131:035001.
243. Nicosia RF, Tchao R, Leighton J. Histotypic angiogenesis in vitro: light microscopic, ultrastructural, and radioautographic studies. In Vitro. 1982;18:538–49.
244. Nicosia RF, McCormick JF, Bielunas J. The formation of endothelial webs and channels in plasma clot culture. Scan Electron Microsc. 1984;(Pt 2):793–9.
245. Nicosia RF. Angiogenesis and the formation of lymphaticlike channels in cultures of thoracic duct. In Vitro Cell Dev Biol. 1987;23:167–74.
246. Bonanno E, Iurlaro M, Madri JA, Nicosia RF. Type IV collagen modulates angiogenesis and neovessel survival in the rat aorta model. In Vitro Cell Dev Biol Anim. 2000;36:336–40.
247. Zhu W-H, Guo X, Villaschi S, Nicosia RF. Regulation of vascular growth and regression by matrix metalloproteinases in the rat aorta model of angiogenesis. Lab Invest. 2000;80:545–55.
248. Nicosia RF, Zhu WH, Fogel E, Howson KM, Aplin AC. A new ex vivo model to study venous angiogenesis and arterio-venous anastomosis formation. J Vasc Res. 2005;42:111–19.
249. Aplin AC, Fogel E, Zorzi P, Nicosia RF. The aortic ring model of angiogenesis. Methods Enzymol. 2008;443:119–36.
250. Clark RA, Tonnesen MG, Gailit J, Cheresh DA. Transient functional expression of alphaV-beta 3 on vascular cells during wound repair. Am J Pathol. 1996;148:1407–21.

251. Heydarkhan-Hagvall S, et al. Human adipose stem cells: a potential cell source for cardiovascular tissue engineering. Cells Tissues Organs. 2008;187:263–74.
252. Song Y, Feijen J, Grijpma DW, Poot AA. Tissue engineering of small-diameter vascular grafts: a literature review. Clin Hemorheol Microcirc. 2011;49:357–74.
253. Rnjak-Kovacina J, et al. Tailoring the porosity and pore size of electrospun synthetic human elastin scaffolds for dermal tissue engineering. Biomaterials. 2011;32:6729–36.
254. Michael S, et al. Tissue engineered skin substitutes created by laser-assisted bioprinting form skin-like structures in the dorsal skin fold chamber in mice. PLoS One. 2013;8:e57741.
255. Shimizu T, et al. Fabrication of pulsatile cardiac tissue grafts using a novel 3-dimensional cell sheet manipulation technique and temperature-responsive cell culture surfaces. Circ Res. 2002;90:e40.
256. Shimizu T, et al. Polysurgery of cell sheet grafts overcomes diffusion limits to produce thick, vascularized myocardial tissues. FASEB J. 2006;20:708–10.
257. Rayatpisheh S, et al. Combining cell sheet technology and electrospun scaffolding for engineered tubular, aligned, and contractile blood vessels. Biomaterials. 2014;35:2713–9.
258. Jeong B, Lee KM, Gutowska A, An YH. Thermogelling biodegradable copolymer aqueous solutions for injectable protein delivery and tissue engineering. Biomacromolecules. 2002;3:865–8.
259. Tarasevich BJ, Gutowska A, Li XS, Jeong BM. The effect of polymer composition on the gelation behavior of PLGA-g-PEG biodegradable thermoreversible gels. J Biomed Mater Res A. 2009;89:248–54.
260. Lin G, Cosimbescu L, Karin NJ, Tarasevich BJ. Injectable and thermosensitive PLGA-g-PEG hydrogels containing hydroxyapatite: preparation, characterization and in vitro release behavior. Biomed Mater. 2012;7:024107.
261. Cui X, Breitenkamp K, Finn MG, Lotz M, D'Lima DD. Direct human cartilage repair using three-dimensional bioprinting technology. Tissue Eng Part A. 2012;18:1304–12.
262. Bertassoni LE, et al. Hydrogel bioprinted microchannel networks for vascularization of tissue engineering constructs. Lab Chip. 2014;14:2202–11.
263. Hinderer S, et al. Engineering of a bio-functionalized hybrid off-the-shelf heart valve. Biomaterials. 2014;35:2130–9.
264. Alexander A, Ajazuddin, Khan, J, Saraf S, Saraf S. Poly(ethylene glycol)-poly(lactic-co-glycolic acid) based thermosensitive injectable hydrogels for biomedical applications. J Control Release. 2013;172:715–29.
265. Fu Q, Saiz E, Tomsia AP. Direct ink writing of highly porous and strong glass scaffolds for load-bearing bone defects repair and regeneration. Acta Biomater. 2011;7:3547–54.
266. Kondo A, Xu H, Abe H, Naito M. Thermoresponsive gelling behavior of concentrated alumina suspensions containing poly(acrylic acid) and PEO-PPO-PEO copolymer. J Colloid Interface Sci. 2012;373:20–6.
267. Parsa S, Gupta M, Loizeau F, Cheung KC. Effects of surfactant and gentle agitation on inkjet dispensing of living cells. Biofabrication. 2010;2:025003.
268. Ruardij TG, van den Boogaart MA, Rutten WL. Adhesion and growth of electrically active cortical neurons on polyethylenimine patterns microprinted onto PEO-PPO-PEO triblock-copolymer-coated hydrophobic surfaces. IEEE Trans Nanobioscience. 2002;1:4–11.
269. Ramadhani N, Shabir M, McConville C. Preparation and characterisation of Kolliphor(R) P 188 and P 237 solid dispersion oral tablets containing the poorly water soluble drug disulfiram. Int J Pharm. 2014;475:514–22.
270. Cai ZH, Shi ZQ, O'Shea GM, Sun AM. Microencapsulated hepatocytes for bioartificial liver support. Artif Organs. 1988;12:388–93.
271. Estes BT, Wu AW, Storms RW, Guilak F. Extended passaging, but not aldehyde dehydrogenase activity, increases the chondrogenic potential of human adipose-derived adult stem cells. J Cell Physiol. 2006;209:987–95.
272. Takei T, Sakai S, Yokonuma T, Ijima H, Kawakami K. Fabrication of artificial endothelialized tubes with predetermined three-dimensional configuration from flexible cell-enclosing alginate fibers. Biotechnol Prog. 2007;23:182–6.

273. Eslaminejad MB, Mirzadeh H, Mohamadi Y, Nickmahzar A. Bone differentiation of marrow-derived mesenchymal stem cells using beta-tricalcium phosphate-alginate-gelatin hybrid scaffolds. J Tissue Eng Regen Med. 2007;1:417–24.
274. de Guzman RC, Ereifej ES, Broadrick KM, Rogers RA, VandeVord PJ. Alginate-matrigel microencapsulated schwann cells for inducible secretion of glial cell line derived neurotrophic factor. J Microencapsul. 2008;25:487–98.
275. Michalopoulos E, et al. Development of methods for studying the differentiation of human mesenchymal stem cells under cyclic compressive strain. Tissue Eng Part C Methods. 2012;18:252–62.
276. Lau TT, Lee LQ, Leong W, Wang DA. Formation of model hepatocellular aggregates in a hydrogel scaffold using degradable genipin crosslinked gelatin microspheres as cell carriers. Biomed Mater. 2012;7:065003.
277. Jeong SH, Lee DW, Kim S, Kim J, Ku B. A study of electrochemical biosensor for analysis of three-dimensional (3D) cell culture. Biosens Bioelectron. 2012;35:128–33.
278. Zheng H, et al. Rotary culture promotes the proliferation of MCF-7 cells encapsulated in three-dimensional collagen-alginate hydrogels via activation of the ERK1/2-MAPK pathway. Biomed Mater. 2012;7:015003.
279. Missirlis D, Kawamura R, Tirelli N, Hubbell JA. Doxorubicin encapsulation and diffusional release from stable, polymeric, hydrogel nanoparticles. Eur J Pharm Sci. 2006;29:120–9.
280. Stoppel WL, et al. Terminal sterilization of alginate hydrogels: efficacy and impact on mechanical properties. J Biomed Mater Res B Appl Biomater. 2014;102:877–84.
281. Jitraruch S, et al. Alginate microencapsulated hepatocytes optimised for transplantation in acute liver failure. Development of a cell culture surface conversion technique using alginate thin film for evaluating effect upon cellular differentiation. PLoS One. 2014;9:e113609.
282. Eccles SA, Court W, Patterson L, Sanderson S. In vitro assays for endothelial cell functions related to angiogenesis: proliferation, motility, tubular differentiation, and proteolysis. Methods Mol Biol. 2009;467:159–81.
283. Colazzo F, Chester AH, Taylor PM, Yacoub MH. Induction of mesenchymal to endothelial transformation of adipose-derived stem cells. J Heart Valve Dis. 2010;19:736–44.
284. Yalvac ME, et al. Isolation and characterization of stem cells derived from human third molar tooth germs of young adults: implications in neo-vascularization, osteo-, adipo- and neurogenesis. Pharmacogenomics J. 2010;10:105–13.
285. Bobis-Wozowicz S, et al. Genetically modified adipose tissue-derived mesenchymal stem cells overexpressing CXCR4 display increased motility, invasiveness, and homing to bone marrow of NOD/SCID mice. Exp Hematol. 2011;39:686-96:e4.
286. Zhang J, et al. Extracellular matrix promotes highly efficient cardiac differentiation of human pluripotent stem cells: the matrix sandwich method. Circ Res. 2012;111:1125–36.
287. Ewald AJ. Isolation of mouse mammary organoids for long-term time-lapse imaging. Cold Spring Harb Protoc. 2013;2013:130–3.
288. Wosnitza M, Hemmrich K, Groger A, Graber S, Pallua N. Plasticity of human adipose stem cells to perform adipogenic and endothelial differentiation. Differentiation. 2007;75:12–23.
289. Di Felice V, et al. HSP90 and eNOS partially co-localize and change cellular localization in relation to different ECM components in 2D and 3D cultures of adult rat cardiomyocytes. Biol Cell. 2007;99:689–99.
290. Fukushima K, et al. Gene expression profiles by microarray analysis during matrigel-induced tube formation in a human extravillous trophoblast cell line: comparison with endothelial cells. Placenta. 2008;29:898–904.
291. Andersen DC, Jensen L, Schroder HD, Jensen CH. "The preadipocyte factor" DLK1 marks adult mouse adipose tissue residing vascular cells that lack in vitro adipogenic differentiation potential. FEBS Lett. 2009;583:2947–53.
292. Ning H, et al. Identification of an aberrant cell line among human adipose tissue-derived stem cell isolates. Differentiation. 2009;77:172–80.
293. Hendrix MJ, Seftor EA, Seftor RE, Fidler IJ. A simple quantitative assay for studying the invasive potential of high and low human metastatic variants. Cancer Lett. 1987;38:137–47.

294. Lawley TJ, Kubota Y. Induction of morphologic differentiation of endothelial cells in culture. J Invest Dermatol. 1989;93:59S–61S.
295. Passaniti A, et al. A simple, quantitative method for assessing angiogenesis and antiangiogenic agents using reconstituted basement membrane, heparin, and fibroblast growth factor. Lab Invest. 1992;67:519–28.
296. Davis GE, Camarillo CW. Regulation of endothelial cell morphogenesis by integrins, mechanical forces, and matrix guidance pathways. Exp Cell Res. 1995;216:113–23.
297. Jarrell BE, et al. Use of freshly isolated capillary endothelial cells for the immediate establishment of a monolayer on a vascular graft at surgery. Surgery. 1986;100:392–9.
298. Jarrell BE, et al. Use of an endothelial monolayer on a vascular graft prior to implantation. Temporal dynamics and compatibility with the operating room. Ann Surg. 1986;203:671–8.
299. Park PK, et al. Thrombus-free, human endothelial surface in the midregion of a Dacron vascular graft in the splanchnic venous circuit–observations after nine months of implantation. J Vasc Surg. 1990;11:468–75.
300. Williams SK, Kosnik P, Kleinert LB, Vossman E, Lye K. Adipose stromal vascular fraction cells isolated using an automated point of care system improve the patency of ePTFE vascular grafts. Tissue Eng Part A. 2013.
301. Mironov V, et al. Organ printing: tissue spheroids as building blocks. Biomaterials. 2009;30:2164–74.
302. Whatley BR, Li X, Zhang N, Wen X. Magnetic-directed patterning of cell spheroids. J Biomed Mater Res A. 2013.
303. Schiele NR, et al. Laser-based direct-write techniques for cell printing. Biofabrication. 2010;2:032001.
304. Hribar KC, Soman P, Warner J, Chung P, Chen S. Light-assisted direct-write of 3D functional biomaterials. Lab Chip. 2014;14:268–75.
305. Sabatini DD, Bensch K, Barrnett RJ. Cytochemistry and electron microscopy. The preservation of cellular ultrastructure and enzymatic activity by aldehyde fixation. J Cell Biol. 1963;17:19–58.
306. Broom ND, Thomson FJ. Influence of fixation conditions on the performance of glutaraldehyde-treated porcine aortic valves: towards a more scientific basis. Thorax. 1979;34:166–76.
307. Smith RM, Jarett L. Quantitative ultrastructural analysis of receptor-mediated insulin uptake into adipocytes. J Cell Physiol. 1983;115:199–207.
308. Nimni ME, Cheung D, Strates B, Kodama M, Sheikh K. Chemically modified collagen: a natural biomaterial for tissue replacement. J Biomed Mater Res. 1987;21:741–71.
309. Hoch J, Dryjski M, Jarrell BE, Carabasi RA, Williams SK. In vitro endothelialization of an aldehyde-stabilized native vessel. J Surg Res. 1988;44:545–54.
310. Zilla P, et al. Glutaraldehyde detoxification of aortic wall tissue: a promising perspective for emerging bioprosthetic valve concepts [see comments]. J Heart Valve Dis. 1997;6:510–20.
311. Schmidt CE, Baier JM. Acellular vascular tissues: natural biomaterials for tissue repair and tissue engineering Biomaterials. 2000;21:2215–31.
312. Sung HW, Huang DM, Chang WH, Huang RN, Hsu JC. Evaluation of gelatin hydrogel crosslinked with various crosslinking agents as bioadhesives: in vitro study. J Biomed Mater Res. 1999;46:520–30.
313. Mercuri JJ, Lovekamp JJ, Simionescu DT, Vyavahare NR. Glycosaminoglycan-targeted fixation for improved bioprosthetic heart valve stabilization. Biomaterials. 2007;28:496–503.
314. Gamboa-Martinez TC, Luque-Guillen V, Gonzalez-Garcia C, Gomez Ribelles JL, Gallego-Ferrer G. Crosslinked fibrin gels for tissue engineering: two approaches to improve their properties. J Biomed Mater Res A. 2014.
315. Hald ES, Steucke KE, Reeves JA, Win Z, Alford PW. Long-term vascular contractility assay using genipin-modified muscular thin films. Biofabrication. 2014;6:045005.
316. Mercurio AM, Shaw LM. Laminin binding proteins. Bioessays. 1991;13:469–73.
317. Girardot JM, Girardot MN. Amide cross-linking: an alternative to glutaraldehyde fixation. J Heart Valve Dis. 1996;5:518–25.

318. Hendriks M, Everaerts F, Verhoeven M. Alternative fixation of bioprostheses. J Long Term Eff Med Implants. 2001;11:163–83.
319. Masurovsky EB, Peterson ER. Photo-reconstituted collagen gel for tissue culture substrates. Exp Cell Res. 1973;76:447–8.
320. Brauer GM. Properties of sealants containing bis-GMA and various diluents. J Dent Res. 1978;57:597–607.
321. Fukui S, Sonomoto K, Itoh N, Tanaka A. Several novel methods for immobilization of enzymes, microbial cells and organelles. Biochimie. 1980;62:381–6.
322. Fedorovich NE, et al. The effect of photopolymerization on stem cells embedded in hydrogels. Biomaterials. 2009;30:344–53.
323. Almeida JF, Ferreira P, Lopes A, Gil MH. Photocrosslinkable biodegradable responsive hydrogels as drug delivery systems. Int J Biol Macromol. 2011;49:948–54.
324. Buwalda SJ, et al. Self-assembly and photo-cross-linking of eight-armed PEG-PTMC star block copolymers. Biomacromolecules. 2011;12:2746–54.
325. Bahney CS, et al. Visible light photoinitiation of mesenchymal stem cell-laden bioresponsive hydrogels. Eur Cell Mater. 2011;22:43–55; discussion 55.
326. Lin H, Cheng AW, Alexander PG, Beck AM, Tuan RS. Cartilage tissue engineering application of injectable gelatin hydrogel with in situ visible-light-activated gelation capability in both air and aqueous solution. Tissue Eng Part A. 2014;20:2402–11.
327. Gao G, Yonezawa T, Hubbell K, Dai G, Cui X. Inkjet-bioprinted acrylated peptides and PEG hydrogel with human mesenchymal stem cells promote robust bone and cartilage formation with minimal printhead clogging. Biotechnol J. 2015.
328. Van Nieuwenhove I, et al. Photo-crosslinkable biopolymers targeting stem cell adhesion and proliferation: the case study of gelatin and starch-based IPNs. J Mater Sci. 2015;26:5424.
329. Massia SP, Hubbell JA. Covalent surface immobilization of Arg-Gly-Asp- and Tyr-Ile-Gly-Ser-Arg-containing peptides to obtain well-defined cell-adhesive substrates. Anal Biochem. 1990;187:292–301.
330. Drumheller PD, Hubbell JA. Polymer networks with grafted cell adhesion peptides for highly biospecific cell adhesive substrates. Anal Biochem. 1994;222:380–8.
331. West JL, Hubbell JA. Comparison of covalently and physically cross-linked polyethylene glycol-based hydrogels for the prevention of postoperative adhesions in a rat model. Biomaterials. 1995;16:1153–6.
332. Sawhney AS, Pathak CP, Hubbell JA. Interfacial photopolymerization of poly(ethylene glycol)-based hydrogels upon alginate-poly(l-lysine) microcapsules for enhanced biocompatibility. Biomaterials. 1993;14:1008–16.
333. Williams SK, Kleinert LB, Hagen KM, Clapper DL. Covalent modification of porous implants using extracellular matrix proteins to accelerate neovascularization. J Biomed Mater Res A. 2006;78:59–65.
334. Williams SK, Kleinert LB, Patula-Steinbrenner V. Accelerated neovascularization and endothelialization of vascular grafts promoted by covalently bound laminin type 1. J Biomed Mater Res A. 2011;99:67–73.
335. Gorovits BM, Osipov AP, Egorov AM. New immunoassay technique using antibody immobilized on a membrane and a flow cuvette as reaction vessel. J Immunol Methods. 1993;157:11–7.
336. Massia SP, Stark J. Immobilized RGD peptides on surface-grafted dextran promote biospecific cell attachment. J Biomed Mater Res. 2001;56:390–9.
337. Derby B. Printing and prototyping of tissues and scaffolds. Science. 2012;338:921–6.
338. Randolph LD, et al. Ultra-fast light-curing resin composite with increased conversion and reduced monomer elution. Dent Mater. 2014;30:594–604.
339. Lemons J, Natiella J. Biomaterials, biocompatibility, and peri-implant considerations. Dent Clin North Am. 1986;30:3–23.
340. Liu Y, Wang Y. Effect of proanthocyanidins and photo-initiators on photo-polymerization of a dental adhesive. J Dent. 2013;41:71–9.

341. Williams CG, Malik AN, Kim TK, Manson PN, Elisseeff JH. Variable cytocompatibility of six cell lines with photoinitiators used for polymerizing hydrogels and cell encapsulation. Biomaterials. 2005;26:1211–8.
342. Mironi-Harpaz I, Wang DY, Venkatraman S, Seliktar D. Photopolymerization of cell-encapsulating hydrogels: crosslinking efficiency versus cytotoxicity. Acta Biomater. 2012;8:1838–48.
343. Zhong C, Wu J, Reinhart-King CA, Chu CC. Synthesis, characterization and cytotoxicity of photo-crosslinked maleic chitosan-polyethylene glycol diacrylate hybrid hydrogels. Acta Biomater. 2010;6:3908–18.
344. Goering RV, Pattee PA. Mutants of Staphylococcus aureus with increased sensitivity to ultraviolet radiation. J Bacteriol. 1971;106:157–61.
345. Humphrey RM, Meyn RE. Recovery and repair in Chinese hamster cells following UV-irradiation. Johns Hopkins Med J Suppl. 1972;(1):159–67.
346. Randtke AS, Williams JR, Little JB. Selection of mutant human cells whose progression into DNA synthesis is sensitive to UV irradiation. Exp Cell Res. 1972;70:360–4.
347. Morrow J, Prickett MS, Fritz S, Vernick D, Deen D. Mutagenesis studies on cultured mammalian cells. The sensitivity of the asparagine-requiring phenotype to several chemical agents. Mutat Res. 1976;34:481–8.
348. Bryant SJ, Nuttelman CR, Anseth KS. Cytocompatibility of UV and visible light photoinitiating systems on cultured NIH/3T3 fibroblasts in vitro. J Biomater Sci Polym Ed. 2000;11:439–57.
349. Sanches-Silva A. et al. Study of the migration of photoinitiators used in printed food-packaging materials into food simulants. J Agric Food Chem. 2009;57:9516–23.
350. Yano S, et al. Preparation of photocrosslinked fish elastin polypeptide/microfibrillated cellulose composite gels with elastic properties for biomaterial applications. Marine drugs. 2015;13:338–53.
351. Larsen EK, Mikkelsen MB, Larsen NB. Protein and cell patterning in closed polymer channels by photoimmobilizing proteins on photografted poly(ethylene glycol) diacrylate. Biomicrofluidics. 2014;8:064127.
352. Selimovic S, Oh J, Bae H, Dokmeci M, Khademhosseini A. Microscale Strategies for Generating Cell-Encapsulating Hydrogels. Polymers. 2012;4:1554.
353. Dulay MT, Choi HN, Zare RN. Visible light-induced photopolymerization of an in situ macroporous sol-gel monolith. J Sep Sci. 2007;30:2979–85.
354. Okino H, Manabe T, Tanaka M, Matsuda T. Novel therapeutic strategy for prevention of malignant tumor recurrence after surgery: local delivery and prolonged release of adenovirus immobilized in photocured, tissue-adhesive gelatinous matrix. J Biomed Mater Res A. 2003;66:643–51.
355. Lin CC, Ki CS, Shih H. Thiol-norbornene photo-click hydrogels for tissue engineering applications. J Appl Polym Sci. 2015;132.
356. Sharkawy AA, Klitzman B, Truskey GA, Reichert WM. Engineering the tissue which encapsulates subcutaneous implants. I. Diffusion properties. J Biomed Mater Res. 1997;37:401–12.
357. Huang X, et al. Microcapsules embedded with three-dimensional fibrous scaffolds for cell culture and tissue engineering. Tissue Eng Part C Methods. 2010;16:1023–32.
358. Bierwolf J, et al. Primary human hepatocytes from metabolic-disordered children recreate highly differentiated liver-tissue-like spheroids on alginate scaffolds. Tissue Eng Part A. 2012;18:1443–53.
359. Rimann M, Bono E, Annaheim H, Bleisch M, Graf-Hausner U. Standardized 3D Bioprinting of Soft Tissue Models with Human Primary Cells. J Lab Autom. (2015).
360. Fu A, Gwon K, Kim M, Tae G, Kornfield JA. Visible-light-initiated thiol-acrylate photopolymerization of heparin-based hydrogels. Biomacromolecules. 2015;16:497–506.
361. Arakawa C, et al. Photopolymerizable chitosan-collagen hydrogels for bone tissue engineering. J Tissue Eng Regen Med. 2014.

362. Mathes DW, et al. Split tolerance to a composite tissue allograft in a swine model. Transplantation. 2003;75:25–31.
363. Son TI, et al. Visible light-induced crosslinkable gelatin. Acta Biomater. 2010;6:4005–10.
364. Turner AE, Flynn LE. Design and characterization of tissue-specific extracellular matrix-derived microcarriers. Tissue Eng Part C Methods. 2012;18:186–97.
365. Bhattacharya V, et al. Enhanced endothelialization and microvessel formation in polyester grafts seeded with CD34(+) bone marrow cells. Blood. 2000;95:581–5.
366. Sankar S, Rajalakshmi T. Application of poly ethylene glycol hydrogel to overcome latex urinary catheter related problems. Biofactors. 2007;30:217–25.
367. Chou AI, Akintoye SO, Nicoll SB. Photo-crosslinked alginate hydrogels support enhanced matrix accumulation by nucleus pulposus cells in vivo. Osteoarthritis Cartilage. 2009;17:1377–84.

3D Bioprinting and 3D Imaging for Stem Cell Engineering

Vivian K. Lee, Andrew Dias, Mehmet S. Ozturk, Kathleen Chen, Brad Tricomi, David T. Corr, Xavier Intes and Guohao Dai

1 Introduction

Three-dimensional (3D) bio-printing, a rapid prototyping method to construct complex 3D structures through a layer-by-layer approach, allows depositing various types of cells and scaffold materials in the desired 3D pattern, and thus has a great potential in cell and tissue engineering applications. An important advantage of this technique is its capability to simultaneously deposit live cells and biochemical molecules (*e.g.* growth factors) along with biomaterial scaffolds at the desired location to mimic the native tissue architecture or to create a specially-designed 3D micro-environment.

G. Dai (✉) · V. K. Lee · A. Dias · M. S. Ozturk · K. Chen · B. Tricomi · D. T. Corr · X. Intes
Department of Biomedical Engineering, Rensselaer Polytechnic Institute, 110 8th Street, 12180 Troy, NY, USA
e-mail: leev4@rpi.edu

V. K. Lee
e-mail: leev4@rpi.edu

A. Dias
e-mail: diasa@rpi.edu

M. S. Ozturk
e-mail: ozturm@rpi.edu

B. Tricomi
e-mail: tricob@rpi.edu

D. T. Corr
e-mail: CORRD@rpi.edu

X. Intes
e-mail: intesx@rpi.edu

K. Turksen (ed.), *Bioprinting in Regenerative Medicine,* Stem Cell Biology and Regenerative Medicine, DOI 10.1007/978-3-319-21386-6_2

With its flexibility and power, the 3D bio-printing technology has been considered as a versatile tool for controlling stem cell fate and creating stem cell niche. 3D bio-printing system capable of precisely deposit biomaterials in desired 3D pattern, allows fine-adjustment of microenvironment where stem cells are embedded. A variety of bio-printing methodologies have been developed in order to generate different stem cell culture environments (*e.g.* stem cell patterning in single-cell level, controlling embryonic body formation) in an efficient and reproducible way.

Meanwhile, 3D printed structures often incorporate thick opaque scaffold, dense population of cells or cell aggregates. Therefore, there are significant difficulties in visualizing the 3D constructs with current imaging modalities. Biological understanding of stem cell differentiation and function has been mainly achieved in cell culture and tissues via destructive techniques such as western blots, immunohistochemistry or quantitative polymerase chain reaction (qPCR). However, to elucidate the interaction of stem cells with the microenvironment in tissue engineering applications, it is necessary to non-destructively monitor the spatio-temporal proliferation and/or differentiation of these cells in an unperturbed environment. To this end, developing novel molecular imaging techniques is critical to observe stem cell fate, cell-cell interactions, and/or structural features of an engineered tissue [1].

In this chapter, we first review the usage of bio-printing technologies for cell and tissue engineering applications, and discuss the potential of each bio-printing system for stem cell application. We also summarize the usage of commonly-used biomedical imaging modalities to tissue engineering and stem cell applications. Then, as optical imaging modalities are the most widely used imaging tools in stem cell studies, we cover the vast array of optical techniques developed to date. We address advantages and limitations of each bio-printer and imaging system, and suggest a perspective on integrating multiple bio-printing techniques and optical imaging for stem cell and tissue engineering applications.

2 Three-Dimensional Bio-Printing Techniques

2.1 Inkjet-Based Printing

The pioneer bio-printing group initially converted a commercial 2D inkjet printer into a bio-printing by replacing the printer ink with biological ink [2, 3]. While the first bio-printing approach is in use for many tissue engineering applications [4–6], numerous inkjet-based bio-printing systems are newly developed to handle a wide range of biomaterials at increasing resolution and speed. Inkjet-based printers commonly use thermal [7], piezoelectric [8–10], or microvalve-based [11, 12] drop-on-demand methods, in which picoliter or microliter volume of aqueous biological materials are dispensed in drops (Fig. 1a, 1b, 1c).

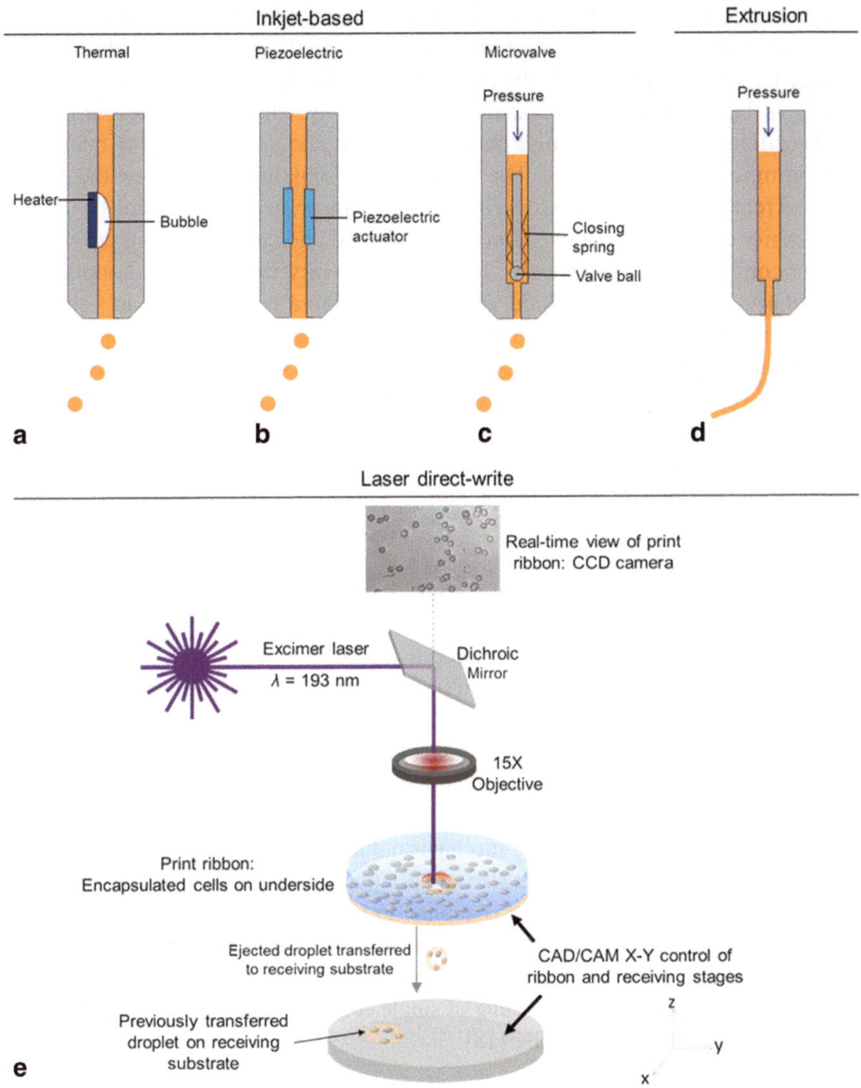

Fig. 1 Bioprinting Techniques. **a–c** Inkjet-based bioprinting technologies, **a** Thermal inkjet printing system, **b** Piezoelectric inkjet printing system, **c** Microvalve-based inkjet printing system, **d** Extrusion-based printing system, **e** Laser direct-write schematic for cell deposition

2.1.1 Printing Mechanism

Thermal bio-printers dispense droplets by increasing temperature of the heating element to produce pulses of pressure (Fig. 1a). The temperature the heater increases for short time, forming a bubble that forces the bioink out of the printer head. Although the temperature of the heating element reaches 200–300 °C during this procedure, it does not have a substantial influence on post-printing viability of mammalian cells [7, 13]. Piezoelectric bio-printing systems utilize piezoelectric actuators or piezocrystals (Fig. 1b). Once electrical signals are applied on the crystals, they generate vibrations inside the printer head to break the bioink into droplets [8, 9, 14]. In both thermal and piezoelectric printing systems, size of nozzle (printer head) orifice and electrical pulse pattern (duration and amplitude) play main roles in determining the droplet size. The viscosity of bioink has an effect on the droplet size and reproducibility, as highly-viscous materials or media suspensions with high cell density often require higher force to be ejected and cause clogging problems.

The components of microvalve-based printers vary by design and application. Commonly-used printers include valve coils, a valve ball, and closing elements in the printer head (Fig. 1c). Unlike thermal or piezoelectric systems, additional pneumatic or piston-derived pressure is applied on the biomaterials loaded inside. The valve coil is triggered by electrical signal, lifting the valve ball, consequently dispensing a droplet. Then, valve orifice is quickly blocked by closing elements [12, 15, 16]. The droplet size is determined by the valve opening duration, the actuation frequency, material viscosity, and the pressure applied.

2.1.2 Resolution and Patterning Capability

Inkjet-based printers create patterns in drop-by-drop manner, in which a series of droplets are closely deposited to form line and surface patterns [17]. Therefore, the resolution of inkjet-based printer is determined by the minimum size of droplet that the printer can generate. The actual resolution of inkjet-based printer tends to be lower than the minimum droplet size because the contour of printed pattern often becomes enlarged by merging of closely printed droplets. Droplet size is depending on numerous parameters, including nozzle diameter, printing mechanism, material viscosity, and substrate properties (e.g. hydrophilicity). Droplet size and printing speed can be controlled electronically, and can range from picoliter to nanoliter in volume with dispensing frequency of 1–10,000 droplets/sec. The spatial resolution of inkjet-based bioprinters ranges from ~50 μm to 1 mm (Table 1). Despite this technique does not support comparable single-cell level of spatial resolution, printing of single cell can be achieved by adjusting cell suspension density and droplet size [18].

Table 1 Comparison of bioprinting techniques

	Inkjet-based [7–12, 18, 81, 149–151]	Extrusion [3, 36, 41, 44–50, 53, 54, 152]	Laser direct-write [58–63, 65, 67, 68, 70, 76–80]
Resolution/droplet size	50 μm–1 mm	5 μm–1 mm	> 10 μm
Material viscosity/density	Low	High	Low–High
(Post-printing) cell viability	High	Medium	High
Single cell control capability	Low	Medium	High
Printing speed (Total fabrication time)	Fast	Slow	Medium-fast

2.1.3 Available Materials

Materials used for bioprinting applications must be biocompatible and cytocompatable, and provide appropriate structural integrity and functional properties during post-printing maturation [19, 20]. Considerations on the viability and functionality of printed bio-structures, limit the range of available chemistries, operating temperature, mechanical and rheological properties. Naturally-derived hydrogel (including collagen, fibrin, chitosan, alginate) and synthetic polymer such as polyethylene glycol (PEG) are commonly-used scaffold materials in the field of tissue engineering [20]. The hydrogel polymers are printed in aqueous precursor form, and then solidified by post-printing crosslinking or gelation process (including enzymatic crosslinking, photo-crosslinking, pH- or temperature-sensitive phase transition). Naturally-derived hydrogels have advantages in supporting biological functions. On the other hand, synthetic polymers are more beneficial for functionalization or tailoring of scaffold material, cost efficiency, and reproducibility.

Hydrogel polymers serve as scaffolds to support 3D structure created via layer-by-layer approach. Thus, the hydrogels used for 3D bioprinting applications are required to have proper mechanical strength in order to maintain the structural integrity during and after printing procedure. The required mechanical strengths often correlate to the concentration and viscosity of polymer. Higher concentration of hydrogel provides more sturdy structure, increased stiffness, and in some cases, better resolution or patterning capability. However, this condition may not be beneficial for embedded cells since they need to degrade the matrix to migrate and proliferate. Hence, material properties need to be specifically optimized for each cell and tissue engineering applications to find balance between structure integrity and preferred cell culture condition.

Inkjet-based printing systems are capable of dispensing most of the hydrogel polymers stated above, but certain materials are not suitable for these printing systems due to the limitation of printing mechanism. The use of gentle ejecting force created by a bubble, vibrations, or pneumatic pressure (< 10 psi) minimizes post-printing cell death, phenotype alteration, or functionality loss, thus has a great advantage in viable cell printing. However, it has a limitation in dispensing highly-viscous materials or high-density cell suspensions that often cause printer head clogging issues and irregular printing patterns. In general, microvalve-based bioprinters have advantages in handling viscos materials compared to the thermal or piezoelectric bioprinters since the additional pressure applied on the loaded materials can be adjusted to create proper ejecting forces.

2.1.4 2D and 3D Cell Printing Applications

Inkjet-based bioprinting systems are more beneficial for dispensing aqueous materials such as cell suspensions (with low cell density) and soluble growth factors. The printing technique can conveniently introduce gradient patterns of cells and growth factors in 3D space by altering droplet size, spacing between droplets, or number of printing times [21–23]. Inkjet-based bioprinters are also widely used for direct cell dispensing to generate 3D *in vitro* co-culture models [16, 24], vascular tissue models [25, 26], and cell aggregates/spheroids [14, 27] maintaining high cell viability.

2.1.5 Advantages and Disadvantages

Advantages of inkjet-based bioprinting technology include wide availability, high cell viability, low cost, and high printing speed. The printing mechanism allows utilizing various biological materials, especially the materials with lower viscosity. The use of gentle pressure guarantees high post-printing viability, showing a great potential of the technique in handling delicate biostructures such as stem cells, progenitor cells, and embryonic bodies [10, 27, 28]. There are some concerns regarding the instant heat exposure in thermal inkjet printers, 15–25 kHz frequencies used in piezoelectric printers, and shear stress caused by microvalve dispensing. These factors may induce cell damages, phenotype alteration, or loss of functionality. However, the strength of these stresses is minimal in comparison with other commonly-used bioprinting technologies. Due to the material viscosity limitation, the aqueous form of hydrogel precursor is widely used for inkjet-based bioprinting platform. Post-printing crosslinking or gelation process is required in this case. These produces include UV irradiation, temperature changes, or the use of acidic/basic solutions, and may induce various cell damages. Another limitation of inkjet-based printers is the difficulties in achieving physiological cell density and matrix density. High concentration of hydrogel polymers or high density cell suspension often cause issues including nozzle clogging, irregular droplet size, irregular dispensing trajectory, and premature gelation.

2.2 Extrusion-Based Printing

Microextrusion is the most commonly-used technique for non-biological 3D printing. A heated printer head extrudes a filament of materials to build 3D shape in additive manufacturing manner. The technique has been successfully applied in bioengineering field, deriving numerous applications in hard tissue replacement/regeneration and porous scaffold designs [29–32].

2.2.1 Printing Mechanism

Extrusion-based bioprinters consist of 3-axis robotic stages and pneumatic [33–36] or mechanical (piston or screw-driven) [37, 38] dispensing system. Continuous pressure is applied on bioink to extrude filaments (Fig. 1d). This technique results continuous line of materials rather than aqueous droplets. Mechanical dispensing systems generally provide more direct control over the material extrusion through the printer nozzle because it can avoid the delay of the gas compression occurred in pneumatic systems. Thus, mechanical dispensing methods are thought to be more suitable for printing highly-viscos materials than pneumatic system. Pneumatic systems support wider range of dispensing pressure thanks to simpler components, whereas mechanical systems that consist of more complex components often had limitations on maximum pressure capabilities.

2.2.2 Resolution and Patterning Capability

The resolution of extrusion-based bioprinter is determined by diameter of extruded filaments. Multiple parameters influence on filament dimension, including nozzle orifice size and geometry, the amount of pressure applied, deposition speed, and mechanical properties of bioink [35, 36, 38–41]. The spatial resolution of extrusion-based bioprinters ranges from ~5 μm to 1 mm, showing its potential in various biomedical applications from single cell deposition to bulk scaffold fabrication (Table 1).

The deposition speed is determined by the moving capability of robotic motors, and is a critical factor to decide total printing time as well as the filament diameter. Current extrusion-based technologies provide 10–50 μm/sec of printing speed [19]. The printing speed could be an obstacle for constructing millimeter- or centi-meter scale of biostructure because it is challenging to maintain cell viability during many hours of printing procedure.

2.2.3 Available Materials

Extrusion-based bioprinting systems support a wide range of viscoelastic properties, with a broad array of biocompatible materials including most of hydrogels described in previous section (2.1.3 Available Materials for inkjet-based bioprinters) and cell spheroids [42]. This type of bioprinters often lose its printing accuracy with low viscos materials as the pressing force of dispensing system may be excessive for those materials [43]. Thus, for direct cell printing, cells are encapsulated within hydrogel to achieve a proper level of viscosity. Materials with shear-thinning properties, which are not suitable for inkjet-based printers, are commonly used for extrusion-based bioprinting platforms.

2.2.4 2D and 3D Cell Printing Applications

Extrusion-based bioprinters have been actively used in the field of tissue engineering, some examples include the fabrication of cardiovascular tissue structure [44–46], *in vitro* multi-layer tissue models [36, 41, 47–49], and 3D cancer co-culture models [50]. The technique has strengths in depositing highly-viscos materials, thus allows achieving physiological cell and matrix density by direct printing. The printed structure with high density experiences higher level of diffusion limit thus requires interconnected hollow structures for sufficient oxygen and nutrient supply. Various bioprinting methodologies have been developed in order to address the perfusion issue including fabrication of porous scaffolds [32], interconnected channels [51], and vascular networks [26, 44, 52].

The ability to print cell- or tissue-spheroid is one of distinctive features of extrusion-based printer. The technique is capable of creating highly-condensed population of cells and matrices and depositing cell spheroids in desired 3D structure for later tissue merging and maturation process [53–55]. The self-assembling spheroid strategy has a potential in tissue organization by direct printing as the multi-cellular spheroid can serve as a biological unit of much complex tissue or organ structures.

2.2.5 Advantages and Disadvantages

The major advantage of extrusion-based bioprinting technique is the ability to dispense high-viscosity and high-density biomaterials. The mechanical properties and cell densities matching with physiological condition is generally beyond the capability of current bioprinting technologies. Extrusion-based bioprinters have a definite advantage in addressing the density issue. Cell-hydrogel mixture with dense cell population, and cell- or tissue-spheroids are deposited through extrusion approach and allowed to fuse and self-assemble into the desired 3D biostructure [46, 53, 54].

Post-printing cell death is the most critical limitation of current extrusion bioprnting. 40–80 % of cell viability [35] is significantly lower than that of inkjet-based printers (> 80 %). The major cell death during the deposition procedure is mainly due to the high dispensing pressure and increase shear stress [35, 39, 56]. The mechanical stresses applied on printed cells may also induce other types of cell damages such as phenotype alterations of stem cells and progenitor cells and loss of cell functionality.

2.3 Laser Direct-Write

2.3.1 Printing Mechanism

Laser Direct-Write (LDW) is a non-contacting method of material deposition that utilizes laser energy absorption to propel a cell-suspended hydrogel droplet to a growth surface. This technique is comprised of two major components: a laser-transparent print ribbon and a receiving substrate. The print ribbon contains both a sacrificial and a transfer layer of material. The laser is pulsed with a configurable energy and repetition rate through the transparent ribbon. The sacrificial layer absorbs the transmitted laser energy, volatizes, and forms a vapor pocket at the ribbon-material interface. This vapor pocket rapidly expands and ejects a droplet of the transfer layer to a receiving substrate. The amount of transferred material can be adjusted with modifications to the laser energy profile. Notably, the rate of mass transfer exceeds the transfer of heat and thereby only negligible amounts of laser thermal energy is transmitted to the deposited transfer material.

For print ribbon preparation, a sacrificial and a transfer layer of material are coated onto the ribbon. The ribbon is first thinly coated with a sacrificial layer. This sacrificial layer will be the only material that will interact with the laser during the transfer step and needs to be able to adhere to the transfer layer material. For recent studies involving transfer of viable cells, a cellular suspension is prepared through the use of cultured mammalian cells resuspended in media or a non-cytotoxic hydrogel. The cellular suspension is then distributed evenly onto the sacrificial layer. The receiving substrate is initially prepared with a hydrogel layer to dampen the kinetic energy of the falling droplet of transferred material. This method allows for the deposition of a high-resolution, 2D pattern of cells, or other bio-payload, on the receiving substrate. A recent study involving alginate deposition and a calcium chloride-coated receiving substrate has also demonstrated *in situ* cross-linking of the hydrogel cellular suspension into a 3D microbead [57]. This envelops cells within a 3D isolated microenvironment, and allows custom placement of cells or other bio-payloads on a planar surface. Typically, this surface is a controlled hydrogel microenvironment, which enables delivery, release, or sequestration of the bio-payload. Direct-written microbead fabrication allows media, growth factors, and waste products to diffuse in and out of each microenvironment on the receiving substrate.

2.3.2 Resolution and Patterning Capability

The LDW system has a camera lens that is coincident with the path of the laser. This setup allows direct visualization of either the transfer layer or receiving substrate. The visualization capability also allows control of which regions of the transfer layer are deposited and placed on the substrate. Furthermore, the size of transfer material that is deposited can be precisely controlled through laser energy adjustment via the beam size. In other bioprinting techniques (e.g. inkjet printing), this step is dictated by the limiting size of the nozzle. For LDW, the printed droplet size is controlled by the selected level of transmitted laser energy. To achieve high spatial patterning resolution, the LDW system has the capacity to independently automate movements of the receiving substrate and ribbon platforms. Thus, highly specific and precise spatially patterning can be achieved and programmed through the use of computer-aided design and computer-aided manufacturing (CAD-CAM) technology. Specifically, high resolution spatial patterning can be attained through the extreme precision and accuracy of the LDW deposition technique, coupled with the CAD-CAM-controlled stages. LDW resolution and patterning capability are therefore within the microscopic tolerance scale for the precise spatial patterning of cells.

2.3.3 Available Materials for LDW

For LDW deposition, a variety of biological materials have been successfully printed. These biomaterials include proteins [58], nucleic acids [59], polymer biomaterials [60], and live cells [61]. To enable material deposition, a sacrificial layer is used to absorb energy from the laser. A vapor pocket forms and ejects the desired material (i.e. transfer material) to the receiving substrate. Sacrificial layers can be created from several different biomaterials such as metals [61, 62] and hygrogels such as Matrigel and gelatin [63, 64]. Each has been successful for multiple cell types. Typical transfer materials include media, glycerol, and various hydrogels (e.g. alginate, gelatin, etc.). Notably for the printing of cells, cytocompatible materials are needed. For the substrate, a large range of materials can be used. However, a softer hydrogel layer is often desirable to cushion cells during bioprinting. Matrigel has been a typical selection for a substrate coating, because it meets this requirement and provides a growth surface [61, 62, 65].

2.3.4 2D Cell Printing Applications

Utilizing the full capability of LDW, live mammalian cells can be deposited in prescribed patterns. LDW has demonstrated with a variety of cell types, including epithelial cells [66], endothelial cells [62, 63, 67, 68], fibroblasts [63, 64] neuroblasts [69], and more. LDW can also be a useful platform in studying stem cells, where spatial control of cell location is needed. Specifically, MSCs, mESCs and hESCs have also been successfully printed (Figs. 2 and 3).

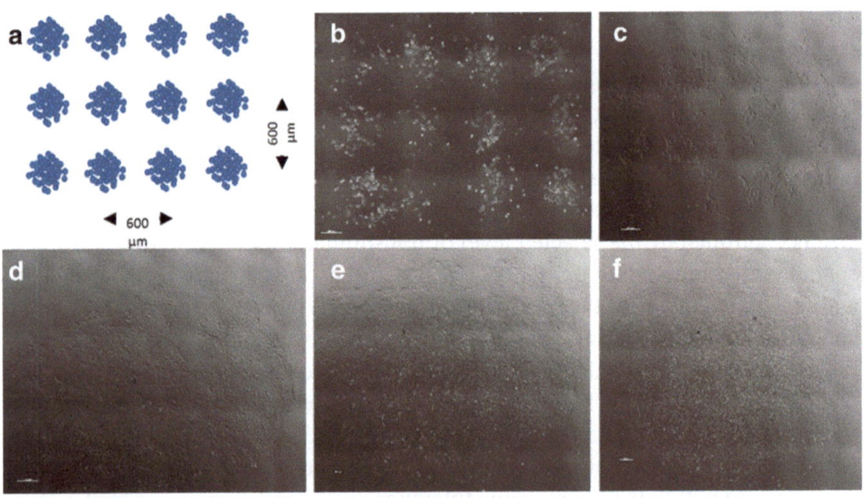

Fig. 2 Grid pattern of hESCs in MEF-CM and bFGF for maintenance of pluripotency, and ROCK inhibitor for survival as single cells, shown **a** schematically and **b–f** under phase contrast microscopy. Cells maintained registry to the initial pattern **b** immediately after printing, but formed a single larger colony over time, showing morphology expected of pluripotent hESCs after **c** 24 h, **d** 48 h, **e** 72 h, and **f** 96 h

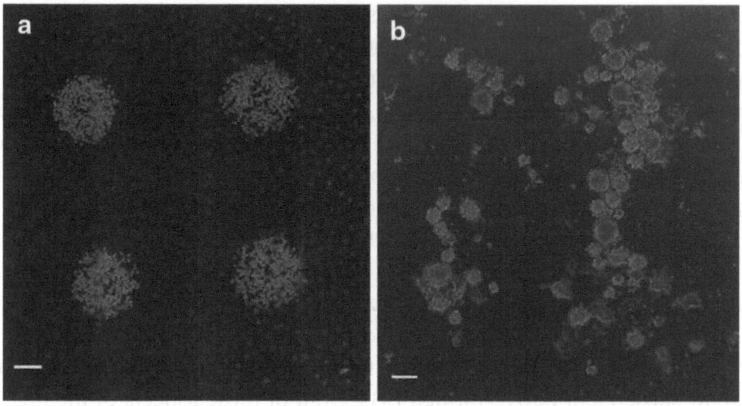

Fig. 3 2 × 2 Grid pattern of CCE mESCs, with 1200 μm spacing, under phase contrast microscopy. Cells maintained registry to the initial pattern, **a** immediately after printing, and formed aggregates and EBs over time, with unrestricted growth, and are shown, **b** after 72 h. Scale bars are 200 μm

Although stem cells have been successfully patterned, the typical approach used is micropatterning or microcontact printing [70–73], rather than LDW or inkjet printing. There are two important differences that distinguishes LDW from these other approaches that may make it particularly attractive for studying stem cells. First, LDW can be used to deposit a pattern of cells on a homogeneous, unpatterned surface. This capability makes LDW fundamentally different from microcontact printing, because microcontact printing is used to pattern islands of adhesive proteins where cells preferentially grow. Cells generally do not migrate or proliferate outside of these adhesion islands, which can be very useful for some studies that seek to limit the size of a cellular colony or maintain a constant distance between cellular colonies. However, adhesion islands do not permit the evolution of a printed structure through migration and proliferation. With adhesion islands, it is also difficult to precisely pattern multiple cell types at discrete locations in a single pattern. For stem cells, an accepted paradigm is that embryonic development allows self-assembly, definition, and evolution of structures *in vivo* during development [74]. To study stem cells, it may be important to allow cells to migrate and self-assemble into structures as they differentiate. It may also be useful to pattern other cell types at precise locations within the stem cell microenvironment to influence their fate.

Secondly, LDW, like inkjet printing, is a non-contact approach. Contact-based approaches (e.g. microcontact printing) are very accurate and precise in deposition, but require high pressures to deposit biomaterials on a planar substrate. Because of the high pressures involved in contact-based methods, cells cannot be directly patterned using these approaches. For cellular patterning, a two-step approach is required: (1) patterning an adhesive protein and (2) subsequent cellular deposition. By contrast, LDW enables direct patterning of cells on a selected homogenous substrate. For stem cell applications, non-contact approaches allow a pattern or configuration of stem cells to be defined as an initial guidance cue. Because growth is unrestricted, "self-patterning" capabilities of stem cells can still be exploited, as the stem cells organize based on the initial patterned state.

The ability of LDW to directly pattern stem cells on a homogeneous substrate, with no restriction on stem cell growth, proliferation, and migration, have been illustrated in the printing of mESCs on a 2D substrate. As mESCs naturally differentiate, 3D spheroids form known as embryoid bodies (EBs). LDW-printed mESCs have shown the ability to form EBs, and outgrowths from these EBs express markers for each of the three germ layers, which strongly suggests pluripotency [75]. LDW can influence EB size, which has been shown to affect differentiation [70, 76]. In printed colonies of controlled size and areal cell density, the cell density influenced EB diameter, but the colony diameter did not. This provides a notable advantage to LDW since it can be used to independently control colony diameter and size of subsequent newly-formed EBs. Both of these aspects of the mESC microenvironment may influence differentiation.

2.3.5 3D Cell Printing Applications

While 2D patterning has shown utility in elucidating cellular behavior and improving understanding of stem cell fate decisions, patterning a 3D microenvironment is closer for mimicking a true *in vivo* microenvironment. LDW offers some capabilities for 3D patterning of both cells and biomaterials. There are at least two different approaches that can be useful for studying stem cells.

MSCs have been successfully printed in 2D [77] and in grid patterns to promote the formation of MSC-derived osteogenic and chondrogenic lineages [78]. The same technique has been also used in a layer-by-layer fashion utilizing fibroblasts and keratinocytes [79], and for endothelial and smooth muscle cells to be printed into 3D scaffolds [80]. For the layer-by-layer approach, cells were suspended in alginate, and printed onto a substrate with a mixture of alginate and blood plasma. Each layer was cross-linked *in situ* by wetting with calcium chloride after printing. Alternatively, cells were suspended in a collagen/media/hydrogen carbonate mixture, and then printed. This approach yielded up to 40 layers, with 500 μm of total thickness. However, this structure exhibited shrinking, which can be expected with collagen gels. Layer-by-layer printing approaches have shown utility for stem cells, since stem cells can be deposited in co-culture with other cell types in a controlled fashion.

Another LDW-based approach for patterning cells in 3D microenvironments involves printing microbeads that encapsulate cells. Using LDW, alginate microbeads were fabricated in a single step by depositing cells suspended in alginate to a calcium chloride/gelatin mixture [57]. The calcium chloride on the substrate cross-linked the alginate into 3D microbeads *in situ*, with excellent control of both microbead size and position, as well as high cell viability. Further applications of this technique include polymer processing of the microbeads for creating hollow shelled structures, or bead-by-bead fabrication for the use of more complex structures. This is especially relevant for stem cells, as printing volume pixels, or "voxels" of stem cells in microbeads enables studying the effects of geometry-based stem cell interactions in a novel fashion.

2.3.6 Advantages and Disadvantages of LDW

LDW has a number of advantages over contact-based approaches, including the ability of LDW to directly pattern cells on a homogeneous planar surface. Of the non-contact cell printing approaches, LDW offers the finest resolution, as it can deposit very small droplets of cells with high-level, potentially sub-10-μm, accuracy [63, 65]. LDW systems can be set up to visualize cells in real time before and after they are deposited, which no other approach offers. This capability ensures specific cells can be chosen for transfer and confirmed visually post-transfer. By contrast, if a smaller number of cells is desired to be printed using an inkjet technique, cells are randomly dispersed in a volume. The number of cells deposited is therefore a function of the probability of the number of cells present in the dispensed volume. Furthermore, because LDW is a nozzle-free printing approach, it may be possible to print a larger range of materials that may otherwise clog a nozzle, such as more viscous hydrogels.

However, LDW may not be appropriate for every application, and its limitations should be considered with other printing approaches. Compared to inkjet printing, LDW has lower throughput, as printing multiple droplets requires movement of the ribbon and receiving stages, and pulsing the laser. The speed of stage movement can limit the rate at which single droplets are deposited. Droplet volume is also generally smaller than droplets printed using inkjet techniques. Smaller droplet volume requires more droplets to cover the same area, and this is also linked to throughput, especially for larger areas.

3 Bio-Printing for Stem Cell Engineering

3.1 Stem Cell Niche

Stem cells have the potential to differentiate into multiple types of cells, which makes them potentially very useful for therapeutic applications. From a single type of cell, it may be possible to generate multiple cell or tissue types. Within the generic term of "stem cells", there are different types of stem cells that may be of use for different applications. Embryonic stem cells (ESCs) are pluripotent. In other words, they can differentiate into any type of somatic cell, from any of the three germ layers. However, differentiation can be very difficult to control, and undesired cell types can often appear. The original cell source is from the inner cell mass of an embryo, so for clinical applications, immune rejection is a concern. There may be a cell type with the same potential as embryonic stem cells, but without the immune complications and ethical controversy of embryonic stem cells. Induced pluripotent stem cells (iPSCs) are generated from a somatic cell source that has been de-differentiated to an embryonic state. It is unclear whether iPSCs are equivalent to ESCs, but their pluripotency has been demonstrated. Finally, there are numerous "adult" stem cells that have multipotency—they can differentiate into multiple cell types, but not every cell type. These types of stem cells include hematopoietic stem cells (HSCs), mesenchymal stem cells (MSCs), and adipose-derived stem cells (ASCs), and have been isolated from various adult tissues.

The stem cell niche is a term for all of the surrounding factors that influence stem cell self-renewal, differentiation, growth, or quiescence. Many aspects of the niche influence cell behavior including mechanical and soluble signals, and these signals can be either external or provided by other cells in the niche. Other cells can provide soluble cues via paracrine signaling, and mechanical cues by physical cell-cell contact. The spatial orientation of stem cells within the niche is potentially very important, because *in vivo*, cells self-organize to produce spatial geometries necessary for development. *In vitro*, self-organization is much more difficult without the cues of the uterus. It may therefore be useful to help cells organize by controlling their spatial location relative to other cells.

3.2 Bioprinting Applications to Influence Stem Cell Signaling and Differentiation

Bioprinting approaches have been employed with stem cells for various ends. While 3D constructs can be achieved with bioprinting, there is also merit in using high-throughput capabilities of bioprinting to generate microstructures, or for cellular studies. Of note, printing has been employed to generate embryoid bodies (EBs) from embryonic stem cells of controlled size in a high-throughput manner. Droplets of cell suspension have been printed to the lid of a Petri dish for embryoid body culture using the hanging drop method [81]. This method was able to produce size-controlled embryoid bodies by varying the cell seeding density, droplet size, and culture time. The variance of EB sizes was much smaller than that of using conventional pipetting, and EBs were generated in a high-throughput manner. Inkjet printing was also used to generate cell concentration and droplet size gradients for hanging drop applications [27].

Control of EB size has been used to influence stem cell differentiation, demonstrated with microwells to control EB size [82–85]. Microcontact printing has also been used to show that colony size influences differentiation [86, 87], as smaller colonies appeared to enrich endoderm populations, while larger colonies appeared to enrich neural populations [70]. This sort of enrichment may be controlled by signaling responses from self-organization of stem cells within colonies [87]. To date, although bioprinting applications have shown excellent control over EB size, they have not yet been used to direct stem cell fate. In addition to control over EB size, bioprinting can be used to independently control stem cell colony size, EB size, and cellular location on a substrate. LDW has been utilized to control EB size, by adjusting printed cell density, as well as colony size location, by precise CAD/CAM X-Y control of the substrate [88]. Altogether, inkjet printing and LDW offer a high degree of control over EB formation, with high throughput, that could be applied for stem cell differentiation studies.

4 3D Imaging Techniques for Cell and Tissue Engineering Applications

Biological understanding of stem cell differentiation and function has been mainly achieved in cell culture and tissues via destructive techniques such as western blots, immunohistochemistry or quantitative polymerase chain reaction. However, to elucidate the interaction of stem cells with the microenvironment in tissue engineering applications, it is necessary to monitor non-destructively the spatio-temporal proliferation and/or differentiation of these cells in an unperturbed environment. To this end, molecular imaging techniques have been developed to observe stem cell fate, interactions between cells, and/or structural features of an engineered tissue [1]. Techniques such as Widefield Fluorescence Microscopy (WFM) and Scanning Electron Microscopy (SEM) have been the mainstay for molecular/cellular activity and surface morphology imaging in 2D cell cultures, transition to a 3D tissue construct requires the use of different imaging modalities due to depth limitations of typical imaging modalities used in 2D cultures. Herein, we first summarize the usage of well-known biomedical imaging modalities (magnetic resonance imaging (MRI), positron emission tomography (PET) and X-ray computed tomography (CT)) to tissue engineering and stem cells applications. Then, as optical imaging modalities are the most widely used imaging tools in stem cell studies, we cover the vast array of optical techniques developed to date. We provide in Table 2 a summary of all imaging modalities discussed in this section and their respective performances. The performances selected summarize general criteria of imaging modalities as well as specific requirements to tissue engineering applications. Of note are bioreactor compatibility and use of labeling agents. As tissue engineering applications aim at mechanistic understanding of stem cell fate, bioreactors have been developed to control precisely the physical environmental parameters, allow for perfusion, and create controlled perturbations. Such bioreactors may not be compatible with the imaging techniques employed. Then imaging sessions should be performed by opening the bioreactor and/or removing the tissue from it. This can lead to perturbations in the microenvironment that can be detrimental to longitudinal investigations.

Also, direct visualization of stem cells in thick tissue based on endogenous contrast is not feasible at this time. Hence, stem cell imaging is performed via labeling of the cells with contrast agents or reporter genes. The ability of molecular imaging modalities to harness such molecular probes and multiplex them for assessing multiple biomarkers simultaneously is of great help in understanding stem cell biology. However, ease of labeling, toxicity, sensitivity and longitudinal stability of the labeling are all parameters of importance when using such labels. Moreover, many of these labeling techniques are restricted to the bench as only a very limited number of molecular probes are approved for clinical use.

Table 2 Comparison of optical imaging techniques

Technique	Resolution[a]	Agent required?	Field of view	BC	Imaging depth	Cost	Acquisition time¬
CM [101, 104, 106, 153, 154]	Lateral: 0.5 μm [155, 156] Axial: 1 μm [155, 156]	Yes	~500 × 500 $μm^2$ [b][157–159]	Yes	100 μm [160, 161]	Low	12 frame/sec [155]
MPM [101, 104, 106, 110, 153]	Lateral: 0.5 μm [101, 104, 154] Axial: 1 μm [101, 104, 154]	Yes	~500 × 500 $μm^2$ [b][162, 163]	Yes	600 μm [164, 165] †	High	1 frame/sec [165]
OCT [101, 104, 153, 154, 166]	Lateral: 19 μm [167, 168] Axial: 3–15 μm [167, 168]	No	2 × 2 mm^2 †[169]	Yes	200 um†,‡[170, 171] 1–3 mm†[172]	Moderate	1 frame/sec (time domain)[168] Max: 7 frame/s (wide field)[168]
PAT [101, 104, 133, 166, 173]	Lateral: 1 μm [101, 104, 166] Spatial: 50–150 μm [101, 104, 166]	No	1–15 mm^2 [136, 174]	Yes	Up to 20 mm	Medium	50 frame/sec [173]
MRI [104, 106, 153, 154]	Spatial: 5–200 μm [156, 175, 176] Lateral: 100 μm [156, 175, 176] Axial: 100 μm [156, 175, 176]	Yes	Variable[c]	No	Whole body	High	5–40 min total (number of frames variable) [175, 176]
CT [104, 106, 153, 154, 166]	Spatial: 5 μm Lateral: 1 μm Axial: 100 μm	Yes	Variable[c]	No	Whole body	High	Slow

Table 2 (continued)

Technique	Resolution[a]	Agent required?	Field of view	BC	Imaging depth	Cost	Acquisition time¬
LOT [122, 177]	100–200 μm	No	10 × 10 mm^2 [125, 178, 179]	Yes	100–200 μm [124, 126, 128, 178, 180]	Moderate	Fast
SPIM [106, 110]	Spatial: 6 μm [110]	Yes	1–5 mm [110]	Yes	1–10 μm [110] 500 μm	Low	Slow
PET [104, 154, 181]	Spatial: 1–2 mm 5 mm [96]	Yes	Variable[c]	No	Whole body	High	Slow
OPT [106, 153]	2–10 μm [182] Spatial: 50–100 μm	Yes	1–10 mm^2 [182]	No	1–3 μm [182] 5–10 mm (if low scattering)	Low	Fast

[a] Resolution depends on numerical aperture and excitation wavelength; [b] Higher resolution reduces due to magnification factor; [c] Limited by the physical space of the device; † Typical value for scattering tissue; ‡ Highly dependent on the scattering coefficient of the tissue; ¬ One frame denotes one field of view

4.1 Biomedical Imaging Modalities

4.1.1 Magnetic Resonance Imaging (MRI)

Magnetic Resonance Imaging (MRI) is well established as an anatomical and functional imaging for clinical and preclinical applications. MRI has been a valuable imaging method for tissue engineering applications [89] due to its non-invasive, high resolution *in vitro* and *in vivo* imaging capabilities. Common MRI sequences are typically used in tissue engineering to generate spatial maps of tissue structure, water diffusion coefficients, and the stiffness of developing engineered tissues. However, for stem cell tracking, direct labeling is required. The most widely used contrast enhancement employed are Gadolinium-based contrast agents or Super Paramagnetic Iron Oxide Nanoparticles (SPIONs)[1, 90]. For instance, cells of interest can be labeled with SPIONs to affect T2 relaxivity [1]. A wide variety of stem cells including Mesenchymal Stem Cells (MSC)[91], neural stem cells [92], human stem cells [93], and smooth muscle cells [94] have been successfully imaged through direct labeling using these paramagnetic particles. The advantage of MRI labeled approaches is that both Gadolinium and superparamagnetic iron oxide nanoparticles have been cleared by the EMA and FDA. Hence, the use of these contrast agents is not limited to the bench and preclinical studies, but can be translated to the clinic [1]. The main limitation of MRI methods for stem cell imaging is their relative low sensitivity. Moreover, SPION-based imaging is not well suited for longitudinal studies as the particles may not stay in the labeled cells. Overall, MRI is a very promising modality for stem cell research from the in vitro stage to the clinical stage. Indeed, MRI offers the potential to track the initial localization of stem cells from tissue engineering construction to transplantation at high resolution.

3.4.2 Positron Emission Tomography (PET)

Positron Emission Tomography (PET) has become the imaging modality of choice for functional and molecular imaging studies. Even though PET is characterized by a relatively low resolution, it offers exquisite specificity and sensitivity. The significance of PET sensitivity (femto-molar concentrations) for tissue engineering and stem cells imaging is that it is an effective method for detection of small cell densities [95]. This sensitivity allows for tracking of stem cells over hours and even days after implantation [1]. The most commonly employed isotope, ^{18}F-Fluoro-deoxygluocse (^{18}F-FDG), allows tracking of cells for 6–8 h in clinical settings [1]. Additionally, cell viability and metabolic activity can be localized with PET by utilizing the glycolic activity of a radionuclide compound, e.g. fluorodeoxyglucose (^{18}FDG).

However, PET suffers from a few limitations. First, PET is a low resolution molecular imaging modality that offers poor signal localization. Thus, in order to improve spatial information, PET needs to be combined with an anatomical imaging modality such as CT or MRI [96]. Second, long exposure time (30–60 min) is required to detect extremely low cell densities [97], which makes PET imaging susceptible to motion artifacts. Lastly, the short lifetimes of radioactive compounds, which require on-site preparation and immediate consumption, hinder the wide use of the technique in stem cell applications. Despite the mentioned limitations and low spatial resolution of PET, it is still the method of choice for clinical translation in stem cell research.

4.1.3 X-ray Computed Tomography (CT)

X-ray Computed Tomography (CT) has been used in both clinical and pre-clinical applications as an anatomical imaging modality for many years. The CT contrast mechanism is based on absorption of X-ray radiation. CT primarily focuses on hard tissue characterization, e.g. bone or mineralized tissue, for *in-vivo* studies [98] as well as *ex-vivo* porosity characterization of engineered tissue [99, 100]. Absorption-based X-ray CT has low contrast for differentiating soft tissue [101] unless freeze-drying procedures are applied or a dry tissue is used. This, of course, puts serious limitations on live tissue examination. An alternative approach is to use metal-based contrast agents (magnetic nanoparticles). It was successfully demonstrated that these types of agents provide strong contrast without applying invasive operations. *In-vitro* studies for stained cells were demonstrated [102, 103] for non-invasive assessment. A combination of whole body imaging and labeling cells is compelling, but toxicity of heavy metal nanoparticles limits the use of CT for longitudinal studies in stem cell studies. Hence, CT is mainly used as a complementary structural imaging modality to complement molecular imaging techniques such as PET or optical imaging.

4.2 Optical Imaging Techniques

4.2.1 Confocal Microscopy (CM)

Confocal microscopy is a ubiquitous imaging modality in tissue engineering and stem cell imaging. Confocal microscopy can be used for tomographic imaging of thick samples less than a few hundred microns due to the ability to produce depth-resolved images. Use of confocal microscopy has helped in the understanding of many complex effects, such as differentiation of mesenchymal stem cells in bioengineered scaffolds [104]. Stains and dyes provide contrast for the images, and can be used to observe the extracellular matrix or proliferation of cells. It is a noninvasive method that allows for visualization of various interactions and processes *in vivo* or *ex vivo*. However, confocal microscopy is limited in imaging a couple of hundreds of microns in depth. Hence, it has very limited utility in imaging *in vitro* thick scaffold and for *in vivo* applications besides epithelial tissue imaging.

4.2.2 Multiphoton/Two-Photon Microscopy (MPM/TPM)

Multiphoton microscopy produces images by scanning a sample with femtosecond pulses of light with wavelengths in the near-infrared region [104–106] to produce two- or three-dimensional images. If multiple photons of a fraction of the wavelength required for fluorescence are absorbed at the same time, then this is equal to the amount of energy required for a single photon of full wavelength [104, 105]. The use of multiple photons results in higher penetration depth (up to ~300 μm in epithelia tissues with native fluorophores [107]), but the resolution degrades as the penetration depth increases due to an increase in scattering events and attenuation of the signal. Besides its increased depth interrogation, multiphoton microscopy is preferred over "single-photon" techniques such as confocal microscopy because of the reduction in photobleaching and damage to the sample, which allows for increased viability of the sample for time-course imaging sessions. If light at peak-wavelength is used, such as during a pulse illumination, however, photo damage to the sample is possible [104, 108]. Photons in the near-infrared (NIR) wavelength region are able to excite exogenous fluorescent probes at greater depth, which allows researchers to analyze structure and function of the tissue *in vivo* [106]. Increased penetration into the tissue allows for imaging of extracellular matrix or molecular factors within cells [104, 109]. Since scattering and absorption are reduced in the NIR range (therapeutic window) these processes can be observed even a few hundreds micron deep in the tissue while still maintaining cellular resolution. This is especially useful in tissue engineering and stem cell applications, since it is necessary to visualize proliferation and differentiation at the molecular level [108, 109].

4.2.3 Selective Plane Illumination Microscopy (SPIM)

Selective plane illumination microscopy (SPIM) creates images by using a thin sheet of laser light to illuminate the sample from the side. The plane of light is oriented orthogonal to the axis of detection, and only a section of the tissue is illuminated at a time [106, 110]. The final image is composed of many images taken by moving the sample in relation to the light plane until the entire sample is imaged. This technique can require greater amounts of time for completion of the image if a large sample is used. Even though only portions of the tissue are illuminated at a time, SPIM is still affected by light scattering as the light propagates through the tissue [106]. Thus, this imaging method results in lower penetration depth into the sample. Benefits of SPIM are reduced photobleaching, a higher signal-to-noise ratio (SNR), and better resolution.

SPIM has recently been used to visualize embryos as well as in time-course experiments. This means that the data can be collected in real time, resulting in "four-dimensional" imaging [111, 112]. This is especially helpful when observing development or cell differentiation. For instance, Swoger *et al.*, were able to trace cell lineages in the embryos of zebra fish throughout development by using dyes to track movement of cells throughout the development process [112]. It was simple to visualize the dyes in the clear embryos, but at times tissue clearance can be necessary to increase the penetration depth into the tissue, making it impossible to image *in vivo* in those cases. Moreover, the geometry of SPIM (perpendicular illumination) makes it difficult to employ it with bio-chambers and also *in vivo* beside developmental biology samples.

4.2.4 Optical Coherence Tomography (OCT)

Optical coherence tomography (OCT), is sometimes referred to as the "optical counterpart" of the ultrasound scan [113]. This technique is based on optical interference of the backscattered signals from the sample with a reference beam [104, 106, 113]. The resolution of OCT is dictated by the low-coherence light of the light source and a few microns resolution can be achieved in tissues as deep as ~1 mm (depending on the nature of the tissue)[104, 106, 114]. OCT is a structural imaging modality that is based on differences in refractive index within the tissues. OCT imaging can be enhanced using polarization contrast (birefringence—collagen imaging), Doppler signals (fluid flow) and contrast agents (gold nanoparticles being the most common). However, OCT is not sensitive to fluorescence signals. Hence, OCT is mainly used in tissue engineering applications instead if real-time monitoring of tissue engineering maturation via structural markers. Since OCT is a reflection-based modality, it can be used *in vitro* on specimens within a bio-chamber and *in vivo*. Clinical OCT systems [115] that allow clinicians to perform real-time *in vivo* imaging are commercially available. These systems are able to visualize anatomical features without the need for contrast agents or preparation of separate samples, which is beneficial to the patients [114–116]. They may be a very useful tool to asses scaffold transplantation and degradation in clinical settings.

4.2.5 Optical Projection Tomography (OPT)

Optical projection tomography (OPT) is an imaging method that can only be used on transparent or thin media. It is based on light trans-illumination over multiple projections through the tissue, and then reconstruction into a three-dimensional image [106, 117, 118] based on an inverse problem. OPT can be used to produce images of structures and function within tissues without altering its morphological characteristics. Typically, the tissue first needs to be subjected to optical clearing, a process by which the tissue is treated with chemical solvents until it is rendered transparent [106]. This decreases but does not eliminate the effects of photon scattering, which results in a limit of penetration depth but higher resolution. However, clearing protocols can be detrimental to the biomaterials employed. Moreover, due to its trans-illumination and multi-view requirements, OPT is not easily applied to tissue within a bioreactor.

Since this method is most effective on clear tissues, OPT has mainly been used in developmental biology to visualize development and gene expression in embryos

[119, 120]. Use of fluorescent stains or dyes allows for comparison of morphology and anatomy at different points in development. Researchers have also been developing a method known as "flow-OPT," which would be used to visualize flow of fluids in transparent tissues [117]. The tissue clearance method is beneficial in these cases, since it would allow for more accurate visualization of the fluid flow or changes in development due to gene expression or blocking within the tissue.

4.2.6 Laminar Optical Tomography (LOT)

Laminar optical tomography (LOT) is a nascent depth-resolved imaging technique [121]. LOT originates from the combination of two well-established techniques: diffuse optical tomography (DOT) and confocal microscopy (CM). Radially spaced detectors are placed with increasing distance from the light source injection point in epi-configuration. As in OPT, LOT is based on an inverse problem, but in which multiple scattering is modeled; this means that there is no need for clearing agents. Different source-detector distances enable projection to different depths within the target tissue (up to 3 mm) [122] with reported resolution of ~200 μm [123]. LOT is able to reconstruct absorption contrast [124] or fluorescence contrast [125, 126]. Although LOT has appeared in different instrumentation and under different names [126–128], the working principle is based on similar phenomena. LOT and variations of LOT perform in epi-configuration (illumination beam and detectors are on the same side of the sample) in non-contact fashion, which makes it a suitable imaging method for tissue engineering applications, even in bioreactors. It has been shown that LOT is a powerful technique to reconstruct fluorescence molecules [125, 129] and reporter genes [126] in vitro as well as hemodynamic response in vivo in the brain [124]. A recent study brings LOT into a new venue by increasing the resolution power below 100 μm thanks to sparsity constraints [130], which will be able to further the use of LOT [131] in stem cell imaging.

4.2.7 Photoacoustic Tomography, Photoacoustic Microscopy (PAT, PAM)

Photoacoustic tomography (PAT) is a multiphysics imaging modality that aims at providing the sensitivity of optical imaging techniques with the resolution of ultrasound imaging. This modality's resolving power can range from the macroscopic level (PAT) down to the microscopic level (Photoacoustic Microscopy, PAM) [132]. PAT capitalizes on the low scattering of sound waves in deep optically scattering media, which enables high resolution imaging that is otherwise very difficult to achieve with traditional optical imaging techniques [132]. Previous studies claimed a resolution as high as 1/200 of the penetration depth into the sample (up to 7 cm) [132]. PAM has been applied successfully to image 3D scaffolds [133]. PAM has been employed to characterized porous scaffold [134], melanin-labeled cells [135] and neovascularization [136]. Despite the capacity of high resolving power, the amount of absorption contrast provided by stem cells is not sufficient for

PAT. Hence nanoparticles are utilized to increase the contrast and specificity [137, 138]. Recently, studies have demonstrated that multimodal approaches using PAT (functional) with ultrasound (structural) imaging offers the possibility to image and track stem cells labeled with gold nanotracers both *in vitro* and *in vivo* [137, 139]. Although the requirements of labeling agents, the necessity for impedance matching with the ultrasound transducer, relatively low sensitivity, and the mitigation of image artefacts are still limitations for PAT, it is emerging as a very promising new modality for tissue engineering and stem cell applications.

4.3 Conclusion

Non-invasive imaging is essential for longitudinal assessment of stem cells. Herein, we surveyed the broad range of techniques; from optical to nuclear imaging modalities, covering all available areas of the electromagnetic spectrum. Choice of spectral bandwidth not only dictates the light source but also dominates the outcome: structural or functional, tissue level or cellular level, superficial depth or whole body imaging. Special attention needs to be paid to select the appropriate technique depending on the type of information that is sought.

Although there is generally not a clear boundary, MRI, OCT, CT (X-ray CT), and PAT delivers structural images while PET, Confocal, LOT, and MPM yield molecular/functional images. As was discussed above, current literature seeks targeted contrast agents for those modalities, which produce inherently structural images. The use of exogenous agents adds different features (molecular, functional, and spectral) to the spatial dimensions.

Functional (molecular) images can be obtained through use of contrast agents. Those agents serve as a beacon for the targeted tissue/cellular activity. The "beacon molecule" varies from radioactive compounds to organic molecules for different imaging modalities. Risk-benefit analysis is the deciding factor for choosing the particular type of agent within an imaging technique. Effective delivery of molecules or compounds to the target location is a challenge and is still an ongoing effort by the research community. The most effective agents are reporter genes. Since they survive in the cell for the entire lifetime of the cell and multiply as the cell proliferates, reporter genes are invaluable tools for stem cell imaging [1, 140].

As a concluding remark, we would like to point out the impact of multimodal imaging, since none of the modalities alone depict a full picture of the biological activity. Combination of different modalities, such as CT-MRI [141, 142], PET-MRI [140, 143, 144], OCT-LOT [131], show great promise. It is important to observe cell differentiation, proliferation, or migration but also it is important to be able to assess the changes or responses on the ECM, implant, or host tissue in general. A full understanding of cell biology can only be understood with this type of comprehensive approach.

5 Integration of Macro- and Micro-Printing, and Optical Imaging for Stem Cell and Tissue Engineering Applications

5.1 Integration of Macro- and Micro-Printing

In this chapter, we have introduced various 3D bio-printing technologies from ink-jet-based bio-printing, the resolution of which is sub-millimeter scale [16, 25, 145], to Laser Direct-Write, which supports single cell level of resolution [64, 146]. Techniques with higher printing resolution (single-cell level and micrometer(μm)-scale) support precise mimicking of tissue or organ microstructures. However, construction of a complete tissue or organ using these printers is nearly impossible mainly due to limitations on manufacturing speed. Some of the techniques capable of single-cell level patterning still require considerable manufacturing time (hours to days) for even a small piece of tissue, and therefore, are incompatible with the fabrication of large structures. On the other hand, the mm- or cm-scale printing technologies have an advantage in creating viable tissue and organs within a suitable time frame. However, the printing resolution of these techniques restricts construction of microstructures, such as microvasculature networks in lung or glomerulus units in kidney tissue, thus limiting their capability on microenvironment control.

Suitable materials for each bio-printing platform also vary depending on printing mechanism and system specification. For example, several inkjet-based printers [6, 25] allow using freshly harvested cells and low-viscosity hydrogels, but high-viscosity materials are not suitable for those platforms. Other printing techniques utilize high pressure (more than 10psi) [35, 145, 147], which guarantee μm-scale resolution using highly-viscous material, but causes significant physical stress on cells during the printing procedure. In the former case, creating a high-stiffness environment for stem cell differentiation into an osteogenic lineage [148] could be challenging, for instance. In the latter case, stem cells may experience high physical stress, which potentially can alter cell fate or cause cell death after printing procedure.

Each technology has its own advantages in specific range of scale and materials, but none can cover the entire range of printing scale and material selections. Combining different printing techniques could be a feasible approach to address these limitations in cell and tissue engineering. Use of two or more printing systems will compensate the manufacturing limitations of each other, enabling the construction of more viable and controlled structure for cell and tissue engineering applications.

One of the potential approaches for integrating different bio-printing technologies will be a sequential printing procedure using multiple printing systems. Millimeter- or centimeter-size structure/scaffold can be created by macro-scale printing system, and then microstructures can be printed using micro-scale printing system. The printing sequence can be altered and repeated until the desired 3D structure is obtained. This approach could be achieved by (1) simply going back and forth between two or more printing systems; or (2) integrating one technique into other systems. For example, inkjet-based printing system described in [16, 26], is capable of installing different types of piezoelectric microvalves which can dispense significantly less volume of materials. The extrusion type of microvalves can be also integrated into the system to enable printing of high-viscosity materials. Using this approach, delicate biomaterials such as stem cells can be printed using low-stress inducing printers while stiffer scaffold construction can be performed by high-stress inducing and high-speed printing platform.

Another potential approach will be creating microstructures in high-throughput manner using microfabrication techniques, and locate the microstructure units in desired 3D pattern using macro-printing technology. Several bioprinting systems are equipped with nozzles with large opening size, and these are capable of printing microbeads or cell aggregates created by micro-bioprinting technologies or other microfabrication techniques. For example, embryonic bodies in various sizes can be obtained using multiple printing techniques [81, 88], then embedded with a desired 3D pattern using bio-printers with large nozzle size. In a more complex approach, sophisticated biological units such as islet of Langerhans in pancreas or liver lobules may be fabricated using micro-patterning technologies, and then assembled into 3D structure using macro-scale bioprinters.

Technical issues may hamper achievement of these suggested approaches, including travel time between two different printing systems. However, these attempts to combine different technology features will enable controlling both microenvironment and macro-architecture for stem cell engineering applications as well as other cell and tissue engineering applications.

5.2 Integration of Bioprinting and Optical Imaging

3D tissue engineering by bioprinting has the potential to generate large structures that may be difficult to image with superficial imaging modalities. In particular, imaging tissue engineered constructs at the millimeter scale or larger is necessary for multi-layer printed structures, and for when a construct is implanted *in vivo*. Three-dimensional imaging of an engineered tissue has a great impact on the assessment of structure and 3D cellular interactions after printing, as well as hypertrophy and hyperplasia. Engineered tissue may be sustained in a bioreactor, which provides features of the *in vivo* environment, such as temperature, fluid flow, and nutrients. However, the bioreactor puts certain limitation on the imaging modality such as depth, non-contact imaging, etc. Therefore, it is necessary to develop new imaging modalities to monitor fluorescence markers (e.g., protein markers, gene reporters, etc.) within thick tissues in real time via non-contact manner.

To date, different imaging modalities have been used within the above-mentioned limitations. Among them, confocal and multiphoton microscopy served with yielding high-resolution images while being limited with maximum detectable depth (up to 3 mm). While these techniques are suitable for fluorescence imaging with high resolution, they cause fluorophores to photo-bleach in a short time, which hinders longitudinal studies. Moreover, this technique may not be suitable for large, multi-layer bioprinted constructs, or to assess a tissue engineered construct after implantation.

To address this issue, recently Laminar Optical Tomography (LOT), also known as Mesoscopic Fluorescence Molecular Tomography (MFMT), was applied on reporter gene imaging. Since LOT does not require a tight focus on the imaging spot, it relieves the excessive excitation on imaging area. At the same time, LOT was shown to be effective at 3-mm depth in scattering media with multiple markers. Therefore, this imaging modality may be useful to assess large, 3D printed structures. Using multiple markers yields both functional information, from different cell types, and structural information. Despite having lower resolution than multiphoton imaging, LOT may be useful for rapid, real-time imaging and analysis of tissue engineered constructs *in vitro* and *in vivo*.

Optical Coherence Tomography (OCT) is another deep tissue imaging technique, widely used for *in vivo* ophthalmology studies, that can produce high-resolution 3D images for tissue with no requirement for sample preparation. Despite its standardized technology, OCT is limited to structural imaging. Cell shape changes or apoptosis may be detected by OCT, but molecular imaging is difficult, because no markers are used. Overall, OCT is the standard tool for characterization of structural property of scattering tissues but not suitable for molecular imaging.

As emphasized above, in order to have a complete picture, a multi-modal approach is strongly recommended. Future development in the multi-modal approaches may prove useful for structural and functional real-time assessment of 3D bioprinted tissues and may become an invaluable tool for tissue engineering applications.

References

1. Rodriguez-porcel M, Wu JC, Gambhir SS. Molecular imaging of stem cells. StemBook; 2009. pp. 1–19.
2. Wilson WC, Boland T. Cell and organ printing 1: protein and cell printers. Anat Rec A Discov Mol Cell Evol Biol. 2003;272:491–6.
3. Boland T, Mironov V, Gutowska A, Roth Ea, Markwald RR. Cell and organ printing 2: fusion of cell aggregates in three-dimensional gels. Anat Rec A Discov Mol Cell Evol Biol. 2003;272:497–502.
4. Boland T, Xu T, Damon B, Cui X. Application of inkjet printing to tissue engineering. Biotechnol J. 2006;1:910–7.
5. Cui X, Boland T, D'Lima DD, Lotz MK. Thermal inkjet printing in tissue engineering and regenerative medicine. Recent Pat Drug Deliv Formul. 2012;6:149–55.
6. Cui X, Boland T. Human microvasculature fabrication using thermal inkjet printing technology. Biomaterials. 2009;30:6221–7.
7. Xu T, Jin J, Gregory C, Hickman JJJ, Boland T. Inkjet printing of viable mammalian cells. Biomaterials. 2005;26:93–9.
8. Saunders RE, Gough JE, Derby B. Delivery of human fibroblast cells by piezoelectric drop-on-demand inkjet printing. Biomaterials. 2008;29:193–203.
9. Hsiehen HB. High throughput bio-printing with individualized piezoelectric ejectors commercial impact at PARC. Science. 2004;(80-.).

10. Kim JD, Choi JS, Kim BS, Chan Choi Y, Cho YW. Piezoelectric inkjet printing of polymers: stem cell patterning on polymer substrates. Polymer (Guildf). 2010;51:2147–54.
11. Lee W, Lee V, Polio S, Keegan P, Lee J-H, Fischer K, Park J-K, Yoo S-S. On-demand three-dimensional freeform fabrication of multi-layered hydrogel scaffold with fluidic channels. Biotechnol Bioeng. 2010;105:1178–86.
12. Demirci U, Montesano G. Cell encapsulating droplet vitrification. Lab Chip. 2007;7:1428–33.
13. Xu T, Gregory Ca, Molnar P, Cui X, Jalota S, Bhaduri SB, Boland T. Viability and electrophysiology of neural cell structures generated by the inkjet printing method. Biomaterials. 2006;27:3580–8.
14. Stachowiak JC, Richmond DL, Li TH, Brochard-Wyart F, Fletcher Da. Inkjet formation of unilamellar lipid vesicles for cell-like encapsulation. Lab Chip. 2009;9:2003–9.
15. Lee W, Debasitis JC, Lee VK, Lee J-H, Fischer K, Edminster K, Park J-K, Yoo S-S. Multi-layered culture of human skin fibroblasts and keratinocytes through three-dimensional freeform fabrication. Biomaterials. 2009;30:1587–95.
16. Lee VK, Singh G, Trasatti JP, Bjornsson C, Tran TN, Xu G, Yoo S-S, Dai G, Karande P. Design and fabrication of human skin by 3D bioprinting. Tissue Eng Part C Methods. 2013;1–44.
17. Stringer J, Derby B. Formation and stability of lines produced by inkjet printing. Langmuir. 2010;26:10365–72.
18. Nakamura M, Kobayashi A, Takagi F, Watanabe A, Hiruma Y, Ohuchi K, Iwasaki Y, Horie M, Morita I, Takatani S. Biocompatible inkjet printing technique for designed seeding of individual living cells. Tissue Eng. 2005;11:1658–66.
19. Murphy SV, Atala A. 3D bioprinting of tissues and organs. Nat Biotechnol. 2014. doi:10.1038/nbt.2958.
20. Murphy SV, Skardal A, Atala A. (2013) Evaluation of hydrogels for bio-printing applications. J Biomed Mater Res–Part A. 101 A:272–84.
21. Campbell PG, Miller ED, Fisher GW, Walker LM, Weiss LE. Engineered spatial patterns of FGF-2 immobilized on fibrin direct cell organization. Biomaterials. 2005;26:6762–70.
22. Phillippi JA, Miller E, Weiss L, Huard J, Waggoner A, Campbell P. Microenvironments engineered by inkjet bioprinting spatially direct adult stem cells toward muscle- and bone-like subpopulations. Stem Cells. 2008;26:127–34.
23. Lee Y-B, Polio S, Lee W, Dai G, Menon L, Carroll RS, Yoo S-S. Bio-printing of collagen and VEGF-releasing fibrin gel scaffolds for neural stem cell culture. Exp Neurol. 2010;223:645–52.
24. Xu T. Complex heterogeneous tissue constructs containing multiple cell types prepared by inkjet printing technology. Biomaterials. 2013;34:130–9.
25. Lee VK, Lanzi AM, Ngo H, Yoo S-S, Vincent Pa, Dai G. Generation of multi-scale vascular network system within 3d hydrogel using 3d bio-printing technology. Cell Mol Bioeng. 2014. doi:10.1007/s12195-014-0340-0.
26. Lee VK, Kim DY, Ngo H, Lee Y, Seo L, Yoo S-S, Vincent Pa, Dai G. Creating perfused functional vascular channels using 3D bio-printing technology. Biomaterials. 2014;35:8092–102.
27. Faulkner-Jones A, Greenhough S, King JA, Gardner J, Courtney A, Shu W. Development of a valve-based cell printer for the formation of human embryonic stem cell spheroid aggregates. Biofabrication. 2013;5:015013.
28. Tasoglu S, Demirci U. Bioprinting for stem cell research. Trends Biotechnol. 2013;31:10–9.
29. Hutmacher DW. Scaffolds in tissue engineering bone and cartilage. Biomaterials. 2000;21:2529–43.
30. Uadkat HV, Hulsman M, Cornelissen K, et al. An algorithm-based topographical biomaterials library to instruct cell fate. Proc Natl Acad Sci U S A. 2012;109:5905–5.
31. Reichert JC, Cipitria a, Epari DR, et al. A tissue engineering solution for segmental defect regeneration in load-bearing long bones. Sci Transl Med. 2012;4:141ra93.

32. Seitz H, Rieder W, Irsen S, Leukers B, Tille C. Three-dimensional printing of porous ceramic scaffolds for bone tissue engineering. J Biomed Mater Res–Part B Appl Biomater. 2005;74:782–8.
33. Chang CC, Boland ED, Williams SK, Hoying JB. Direct-write bioprinting three-dimensional biohybrid systems for future regenerative therapies. J Biomed Mater Res B Appl Biomater. 2011;98:160–70.
34. Fedorovich NE, Swennen I, Girones J, Moroni L, van Blitterswijk CA, Schacht E, Alblas J, Dhert WJ. Evaluation of photocrosslinked Lutrol hydrogel for tissue printing applications. Biomacromolecules. 2009;10:1689–96.
35. Chang R, Nam J, Sun W. Effects of dispensing pressure and nozzle diameter on cell survival from solid freeform fabrication-based direct cell writing. Tissue Eng Part A. 2008;14:41–8.
36. Chang R, Nam J, Sun W. Direct cell writing of 3D microorgan for in vitro pharmacokinetic model. Tissue Eng Part C Methods. 2008;14:157–66.
37. Jakab K, Damon B, Neagu A, Kachurin A, Forgacs G. Three-dimensional tissue constructs built by bioprinting. Biorheology. 2006;43:509–13.
38. Visser J. Biofabrication of multi-material anatomically shaped tissue constructs. Biofabrication. 2013;5:35007.
39. Smith CM. Three-dimensional bioassembly tool for generating viable tissue-engineered constructs. Tissue Eng. 2004;10:1566–76.
40. Khalil S, Nam J, Sun W. Multi-nozzle deposition for construction of 3D biopolymer tissue scaffolds. Rapid Prototyp J. 2005;11:9–17.
41. Khalil S, Sun W. Biopolymer deposition for freeform fabrication of hydrogel tissue constructs. Mater Sci Eng C. 2007;27:469–78.
42. Peltola SM, Melchels FP, Grijpma DW, Kellomaki M. A review of rapid prototyping techniques for tissue engineering purposes. Ann Med. 2008;40:268–80.
43. Jones N. Science in three dimensions: the print revolution. Nat. 2012;487:22–3.
44. Therriault D, White SR, Lewis JA. Chaotic mixing in three-dimensional microvascular networks fabricated by direct-write assembly. Nat Mater. 2003;2:265–71.
45. Duan B, Hockaday LA, Kang KH, Butcher JT. 3D bioprinting of heterogeneous aortic valve conduits with alginate/gelatin hydrogels. J Biomed Mater Res A. 2013;101:1255–64.
46. Norotte C, Marga FS, Niklason LE, Forgacs G. Scaffold-free vascular tissue engineering using bioprinting. Biomaterials. 2009;30:5910–7.
47. Shim J-H, Lee J-S, Kim JY, Cho D-W. Bioprinting of a mechanically enhanced three-dimensional dual cell-laden construct for osteochondral tissue engineering using a multi-head tissue/organ building system. J Micromechanics Microengineering. 2012;22:085014.
48. Liu Tsang V, Chen AA, Cho LM, Jadin KD, Sah RL, DeLong S, West JL, Bhatia SN. Fabrication of 3D hepatic tissues by additive photopatterning of cellular hydrogels. FASEB J. 2007;21:790–801.
49. Yan Y, Wang X, Pan Y, Liu H, Cheng J, Xiong Z, Lin F, Wu R, Zhang R, Lu Q. Fabrication of viable tissue-engineered constructs with 3D cell-assembly technique. Biomaterials. 2005;26:5864–71.
50. Xu F. A three-dimensional in vitro ovarian cancer coculture model using a high-throughput cell patterning platform. Biotechnol J. 2011;6:204–12.
51. Miller JS, Shen CJ, Legant WR, Baranski JD, Blakely BL, Chen CS. Bioactive hydrogels made from step-growth derived PEG-peptide macromers. Biomaterials. 2010;31:3736–43.
52. Miller JS, Stevens KR, Yang MT, et al. Rapid casting of patterned vascular networks for perfusable engineered three-dimensional tissues. Nat Mater. 2012;11:768–74.
53. Mironov V, Visconti RP, Kasyanov V, Forgacs G, Drake CJ, Markwald RR. Organ printing: tissue spheroids as building blocks. Biomaterials. 2009;30:2164–74.
54. Mironov V, Kasyanov V, Markwald RR. Organ printing: from bioprinter to organ biofabrication line. Curr Opin Biotechnol. 2011;22:667–73.
55. Marga F. Toward engineering functional organ modules by additive manufacturing. Biofabrication. 2012;4:22001.

56. Nair K, Gandhi M, Khalil S, Yan KC, Marcolongo M, Barbee K, Sun W. Characterization of cell viability during bioprinting processes. Biotechnol J. 2009;4:1168–77.
57. Kingsley DM, Dias AD, Chrisey DB, Corr DT. Single-step laser-based fabrication and patterning of cell-encapsulated alginate microbeads. Biofabrication. 2013;5:045006.
58. Ringeisen BR, Wu PK, Kim H, Piqué A, Auyeung RYC, Young HD, Chrisey DB, Krizman DB. Picoliter-scale protein microarrays by laser direct write. Biotechnol Prog. 2002;18:1126–9.
59. Colina M, Serra P, Fernández-Pradas JM, Sevilla L, Morenza JL. DNA deposition through laser induced forward transfer. Biosens Bioelectron. 2005;20:1638–42.
60. Chrisey DB, Piqué A, McGill RA, Horwitz JS, Ringeisen BR, Bubb DM, Wu PK. Laser deposition of polymer and biomaterial films. Chem Rev. 2003;103:553–76.
61. Barron JA, Ringeisen BR, Kim H, Spargo BJ, Chrisey DB. Application of laser printing to mammalian cells. Thin Solid Films. 2004;453–454:383–7.
62. Barron JA, Wu P, Ladouceur HD, Ringeisen BR. Biological laser printing: a novel technique for creating heterogeneous 3-dimensional cell patterns. Biomed Microdevices. 2004;6:139–47.
63. Schiele NR, Koppes RA, Corr DT, Ellison KS, Thompson DM, Ligon LA, Lippert TKM, Chrisey DB. Laser direct writing of combinatorial libraries of idealized cellular constructs: biomedical applications. Appl Surf Sci. 2009;255:5444–7.
64. Schiele NR, Chrisey DB, Corr DT. Gelatin-based laser direct-write technique for the precise spatial patterning of cells. Tissue Eng Part C Methods. 2011;17:289–98.
65. Ringeisen BR, Kim H, Barron Ja, Krizman DB, Chrisey DB, Jackman S, Auyeung RYC, Spargo BJ. Laser printing of pluripotent embryonal carcinoma cells. Tissue Eng. 2004;10:483–91.
66. Hopp B, Smausz T, Kresz N, Barna N, Bor Z, Kolozsvári L, Chrisey DB, Szabó A, Nógrádi A. Survival and proliferative ability of various living cell types after laser-induced forward transfer. Tissue Eng. 2005;11:1817–23.
67. Chen C, Barron J, Ringeisen B. Cell patterning without chemical surface modification: cell–cell interactions between printed bovine aortic endothelial cells (BAEC) on a homogeneous cell-adherent hydrogel. Appl Surf Sci. 2006;252:8641–5.
68. Guillemot F, Souquet A, Catros S, et al. High-throughput laser printing of cells and biomaterials for tissue engineering. Acta Biomater. 2010;6:2494–500.
69. Patz TM, Doraiswamy A, Narayan RJ, He W, Zhong Y, Bellamkonda R, Modi R, Chrisey DB. Three-dimensional direct writing of B35 neuronal cells. J Biomed Mater Res. 2006;78:124–30.
70. Bauwens CL, Peerani R, Niebruegge S, Woodhouse KA, Kumacheva E, Husain M, Zandstra PW. Control of human embryonic stem cell colony and aggregate size heterogeneity influences differentiation trajectories. Stem Cells. 2008;26:2300–10.
71. Sun Y, Villa-Diaz LG, Lam RHW, Chen W, Krebsbach PH, Fu J. Mechanics regulates fate decisions of human embryonic stem cells. PLoS ONE. 2012;7:e37178.
72. Tang J, Peng R, Ding J. The regulation of stem cell differentiation by cell-cell contact on micropatterned material surfaces. Biomaterials. 2010;31:2470–6.
73. Song W, Lu H, Kawazoe N, Chen G. Adipogenic differentiation of individual mesenchymal stem cell on different geometric micropatterns. Langmuir. 2011;27:6155–62.
74. Sasai Y. Next-generation regenerative medicine: organogenesis from stem cells in 3D culture. Cell Stem Cell. 2013;12:520–30.
75. Raof NA, Schiele NR, Xie Y, Chrisey DB, Corr DT. The maintenance of pluripotency following laser direct-write of mouse embryonic stem cells. Biomaterials. 2011;32:1802–8.
76. Hwang Y-SYS, Chung BGBG, Ortmann D, Hattori N, Moeller HCH-C, Khademhosseini A. Microwell-mediated control of embryoid body size regulates embryonic stem cell fate via differential expression of WNT5a and WNT11. Proc Natl Acad Sci U S A. 2009;106:16978–83.
77. Koch L, Kuhn S, Sorg H, et al. Laser printing of skin cells and human stem cells. Tissue Eng Part C Methods. 2010;16:847–54.

78. Gruene M, Deiwick A, Koch L, Schlie S, Unger C, Hofmann N, Bernemann I, Glasmacher B, Chichkov B. Laser printing of stem cells for biofabrication of scaffold-free autologous grafts. Tissue Eng Part C Methods. 2010;17:79–87.
79. Koch L, Deiwick A, Schlie S, et al. Skin tissue generation by laser cell printing. Biotechnol Bioeng. 2012;109:1855–63.
80. Ovsianikov A, Gruene M, Pflaum M, Koch L, Maiorana F, Wilhelmi M, Haverich A, Chichkov B. Laser printing of cells into 3D scaffolds. Biofabrication. 2010;2:014104.
81. Xu F, Sridharan B, Wang S, Gurkan UA, Syverud B, Demirci U. Embryonic stem cell bioprinting for uniform and controlled size embryoid body formation. Biomicrofluidics. 2011;5:22207.
82. Zhang Y, Xia Y. Formation of embryoid bodies with controlled sizes and maintained pluripotency in three-dimensional inverse opal scaffolds. Adv Funct Mater. 2012;22:121–9.
83. Sakai Y, Yoshiura Y, Nakazawa K. Embryoid body culture of mouse embryonic stem cells using microwell and micropatterned chips. J Biosci Bioeng. 2011;111:85–91.
84. Mohr J, Zhang J, Azarin S, Soerens A, de Pablo JJ, Thomson JA, Lyons GE, Palacek SP, Kamp TJ. The microwell control of embryoid body size in order to regulate cardiac differentiation of human embryonic stem cells. Biomaterials. 2010;31:1–18.
85. Konno T, Akita K, Kurita K, Ito Y. Formation of embryoid bodies by mouse embryonic stem cells on plastic surfaces. J Biosci Bioeng. 2005;100:88–93.
86. Bauwens CL, Song H, Thavandiran N, Ungrin M, Masse S, Nanthakumar K, Seguin C, Zandstra PW. Geometric control of cardiomyogenic induction in human pluripotent stem cells. Tissue Eng Part A. 2011;17:1901–9.
87. Warmflash A, Sorre B, Etoc F, Siggia ED, Brivanlou AH. Breaking down pluripotency in the porcine embryo reveals both a premature and reticent stem cell state in the inner cell mass and unique expression profiles of the naive and primed stem cell states. Nat Methods. 2014;11:847–54.
88. Dias aD, Unser aM, Xie Y, Chrisey DB, Corr DT. Generating size-controlled embryoid bodies using laser direct-write. Biofabrication. 2014;6:025007.
89. Ferreira HA, Silva H, Rodrigues LM, Grande C. Magnetic resonance imaging a powerful tool for tissue engineering. Biomed Biopharm Res. 2012;9(2):159–65.
90. Berman SC, Walczak P, Bulte JWM. Tracking stem cells using magnetic nanoparticles. Wiley Interdiscip Rev Nanomed Nanobiotechnol. 2011;3(4):343–55.
91. Andreas K, Georgieva R, Ladwig M, Mueller S, Notter M, Sittinger M, Ringe J. Highly efficient magnetic stem cell labeling with citrate-coated superparamagnetic iron oxide nanoparticles for MRI tracking. Biomaterials. 2012;33:4515–25.
92. Burgess A, Ayala-Grosso Ca, Ganguly M, Jordão JF, Aubert I, Hynynen K. Targeted delivery of neural stem cells to the brain using MRI-guided focused ultrasound to disrupt the blood-brain barrier. PLoS ONE. 2011;6:e27877.
93. Tarantal AF, Lee CCI, Batchelder Ca, Christensen JE, Prater D, Cherry SR. Radiolabeling and in vivo imaging of transplanted renal lineages differentiated from human embryonic stem cells in fetal rhesus monkeys. Mol Imaging Biol. 2012;14:197–204.
94. Ramaswamy S, Schornack Pa, Smelko AG, Boronyak SM, Ivanova J, Mayer JE, Sacks MS. Superparamagnetic iron oxide (SPIO) labeling efficiency and subsequent MRI tracking of native cell populations pertinent to pulmonary heart valve tissue engineering studies. NMR Biomed. 2012;25:410–7.
95. Chouinard Ja, Rousseau Ja, Beaudoin J-F, Vermette P, Lecomte R. Positron emission tomography detection of human endothelial cell and fibroblast monolayers: effect of pretreament and cell density on 18FDG uptake. Vasc Cell. 2012;4:5.
96. DiFilippo FP, Patel S, Asosingh K, Erzurum SC. Small-animal imaging using clinical positron emission tomography/computed tomography and super-resolution. Mol Imaging. 2012;11:210–9.
97. Terrovitis J, Lautamäki R, Bonios M, et al. Noninvasive quantification and optimization of acute cell retention by in vivo positron emission tomography after intramyocardial cardiac-derived stem cell delivery. J Am Coll Cardiol. 2009;54:1619–26.

98. Potter K, Sweet DE, Anderson P, Davis GR, Isogai N, Asamura S, Kusuhara H, Landis WJ. Non-destructive studies of tissue-engineered phalanges by magnetic resonance microscopy and X-ray microtomography. Bone. 2006;38:350–8.
99. Zhang Y, Wu C, Friis T, Xiao Y. The osteogenic properties of CaP/silk composite scaffolds. Biomaterials. 2010;31:2848–56.
100. Jones AC, Arns CH, Sheppard AP, Hutmacher DW, Milthorpe BK, Knackstedt Ma. Assessment of bone ingrowth into porous biomaterials using MICRO-CT. Biomaterials. 2007;28:2491–504.
101. Appel Aa, Anastasio Ma, Larson JC, Brey EM. Imaging challenges in biomaterials and tissue engineering. Biomaterials. 2013;34:6615–30.
102. Zehbe R, Goebbels J, Ibold Y, Gross U, Schubert H. Three-dimensional visualization of in vitro cultivated chondrocytes inside porous gelatine scaffolds: a tomographic approach. Acta Biomater. 2010;6:2097–107.
103. Dorsey SM, Lin-Gibson S, Simon CG. X-ray microcomputed tomography for the measurement of cell adhesionand proliferation in polymer scaffolds. Biomaterials. 2009;30:2967–74.
104. Georgakoudi I, Rice WL, Hronik-Tupaj M, Kaplan DL. Optical spectroscopy and imaging for the noninvasive evaluation of engineered tissues. Tissue Eng Part B Rev. 2008;14:321–40.
105. Tadrous PJ. Methods for imaging the structure and function of living tissues and cells: 3. Confocal microscopy and micro-radiology. J Pathol. 2000;191:345–54.
106. Ntziachristos V. Going deeper than microscopy: the optical imaging frontier in biology. Nat. 2010;7:603–14.
107. Durr NJ, Weisspfennig CT, Holfed Ba, Ben-Yakar A. Maximum imaging depth of two-photon autofluorescence microscopy in epithelial tissues. J Biomed Opt. 2011;16:026008.
108. Rubart M. Two-photon microscopy of cells and tissue. Circ Res. 2004;95:1154–66.
109. Higashi T, Watanabe W, Matsunaga S. Application of visualization techniques for cell and tissue engineering. J Biosci Bioeng. 2013;115:122–6.
110. Huisken J, Stainier DYR. Selective plane illumination microscopy techniques in developmental biology. Dev. 2009;136:1963–75.
111. Kaufmann A, Mickoleit M, Weber M, Huisken J. Multilayer mounting enables long-term imaging of zebrafish development in a light sheet microscope. Dev. 2012;139:3242–7.
112. Swoger J, Muzzopappa M, López-Schier H, Sharpe J. 4D retrospective lineage tracing using SPIM for zebrafish organogenesis studies. J Biophotonics. 2011;4:122–34.
113. Tadrous PJ. Methods for imaging the structure and function of living tissues and cells: 1. Optical coherence tomography. J Pathol. 2000;191:115–9.
114. Schmitt JM. Optical coherence tomography (OCT): a review. IEEE J Sel Top Quantum Electron. 1999;5:1205–15.
115. Calantog A, Hallajian L, Nabelsi T, Mansour S, Le A, Epstein J, Wilder-Smith P. A prospective study to assess in vivo optical coherence tomography imaging for early detection of chemotherapy-induced oral mucositis. Lasers Surg Med. 2013;45:22–7.
116. Leitgeb RA. Optical coherence tomography–high resolution imaging of structure and function. Conf Proc Annu Int Conf IEEE Eng Med Biol Soc. 2007;2007:530–2.
117. Bassi A, Fieramonti L, D'Andrea C, Mione M, Valentini G. In vivo label-free three-dimensional imaging of zebrafish vasculature with optical projection tomography. J Biomed Opt. 2011;16:100502.
118. Sharpe J. Optical projection tomography. Annu Rev Biomed Eng. 2004;6:209–28.
119. Sharpe J, Ahlgren U, Perry P, Hill B, Ross A, Hecksher-Sørensen J, Baldock R, Davidson D. Optical projection tomography as a tool for 3D microscopy and gene expression studies. Sci. 2002;296:541–5.
120. Sharpe J. Optical projection tomography as a new tool for studying embryo anatomy. J Anat. 2003;202:175–81.
121. Dunn A, Boas D. Transport-based image reconstruction in turbid media with small source-detector separations. Opt Lett. 2000;25:1777–9.

122. Hillman EMC, Boas DA, Dale AM, Dunn AK. Laminar optical tomography:demonstration of millimeter-scale depth-resolved imaging in turbid media. Opt Lett. 2004;29:1650.
123. Hillman EMCMC, Devor A, Dunn AK, Boas Da. Laminar optical tomography: high-resolution 3D functional imaging of superficial tissues. Proc SPIE. 2006;6143:61431M–61431M–14.
124. Ouakli N, Guevara E, Dubeau S, Beaumont E, Lesage F. Laminar optical tomography of the hemodynamic response in the lumbar spinal cord of rats. Opt Express. 2010;18:10068–77.
125. Zhao L, Lee VK, Yoo S-S, Dai G, Intes X. The integration of 3-D cell printing and mesoscopic fluorescence molecular tomography of vascular constructs within thick hydrogel scaffolds. Biomaterials. 2012;33:5325–32.
126. Ozturk MS, Lee VK, Zhao L, Dai G, Intes X. Mesoscopic fluorescence molecular tomography of reporter genes in bioprinted thick tissue. J Biomed Opt. 2013;18:100501.
127. Björn S, Ntziachristos V, Schulz R. Mesoscopic epifluorescence tomography: reconstruction of superficial and deep fluorescence in highly-scattering media. Opt Express. 2010;18:8422–9.
128. Chen C-W, Chen Y. Optimization of design parameters for fluorescence laminar optical tomography. J Innov Opt Health Sci. 2011;04:309–23.
129. Burgess SA, Yuan B, Bouchard MB, Ratner D, Hillman EMC. Simultaneous multi-wavelength laminar optical tomography imaging of skin cancer. Biomed Opt. 2008;3–5.
130. Yang F, Ozturk M, Zhao L, Cong W, Wang G, Intes X. High-resolution mesoscopic fluorescence molecular tomography based on compressive sensing. IEEE Trans Biomed Eng. 2014. doi:10.1109/TBME.2014.2347284.
131. Chen C-W, Yeatts AB, Fisher JP, Chen Y. (2011) Three-dimensional imaging of stem cell distribution within tissue engineering scaffolds using angled fluorescent laminar optical tomography (aFLOT). IEEE Photonic Soc 24th Annu Meet. 4:729–30.
132. Wang LV, Hu S. Photoacoustic tomography: in vivo imaging from organelles to organs. Sci. 2012;335:1458–62.
133. Cai X, Zhang YS, Xia Y, Wang LV. Photoacoustic microscopy in tissue engineering. Mater Today (Kidlington). 2013;16:67–77.
134. Cai X, Paratala BS, Hu S, Sitharaman B, Wang LV. Multiscale photoacoustic microscopy of single-walled carbon nanotube-incorporated tissue engineering scaffolds. Tissue Eng Part C Methods. 2012;18:310–7.
135. Zhang Y, Cai X, Choi SW, Kim C, Wang LV, Xia Y. Chronic label-free volumetric photoacoustic microscopy of melanoma cells in three-dimensional porous Scaffolds. Biomaterials. 2010;31:8651–8.
136. Cai X, Zhang Y, Li L, Choi S-W, MacEwan MR, Yao J, Kim C, Xia Y, Wang LV. Investigation of neovascularization in three-dimensional porous scaffolds in vivo by a combination of multiscale photoacoustic microscopy and optical coherence tomography. Tissue Eng Part C Methods. 2013;19:196–204.
137. Nam SY, Ricles LM, Suggs LJ, Emelianov SY. In vivo ultrasound and photoacoustic monitoring of mesenchymal stem cells labeled with gold nanotracers. PLoS ONE. 2012;7:e37267.
138. Chung E, Nam SY, Ricles LM, Emelianov SY, Suggs LJ. Evaluation of gold nanotracers to track adipose-derived stem cells in a PEGylated fibrin gel for dermal tissue engineering applications. Int J Nanomed. 2013;8:325–36.
139. Wang B, Karpiouk A, Yeager D, Amirian J, Litovsky S, Smalling R, Emelianov S. In vivo intravascular ultrasound-guided photoacoustic imaging of lipid in plaques using an animal model of atherosclerosis. Ultrasound Med Biol. 2012;38:2098–103.
140. Nguyen PK, Riegler J, Wu JC. Stem cell imaging: from bench to bedside. Cell Stem Cell. 2014;14:431–44.
141. Chou S, Shau Y, Wu P, Yang Y. In vitro and in vivo studies of FePt nanoparticles for dual modal CT/MRI molecular imaging. J Am Chem Soc. 2010;132(38):13270–8.

142. Amado LC, Schuleri KH, Saliaris AP, et al. Multimodality noninvasive imaging demonstrates in vivo cardiac regeneration after mesenchymal stem cell therapy. J Am Coll Cardiol. 2006;48:2116–24.
143. Catana C, Procissi D, Wu Y, Judenhofer MS, Qi J, Pichler BJ, Jacobs RE, Cherry SR. Simultaneous in vivo positron emission tomography and magnetic resonance imaging. Proc Natl Acad Sci U S A. 2008;105:3705–10.
144. Catana C, Drzezga A, Heiss W-D, Rosen BR. PET/MRI for neurologic applications. J Nucl Med. 2012;53:1916–25.
145. Kolesky DB, Truby RL, Gladman aS, Busbee Ta, Homan Ka, Lewis Ja. 3D bioprinting of vascularized, heterogeneous cell-laden tissue constructs. Adv Mater. 2014;26:3124–30.
146. Schiele NR, Corr DT, Huang Y, Raof NA, Xie Y, Chrisey DB. Laser-based direct-write techniques for cell printing. Biofabrication. 2010;2:32001.
147. Smith CM, Christian JJ, Warren WL, Williams SK. Characterizing environmental factors that impact the viability of tissue-engineered constructs fabricated by a direct-write bioassembly tool. Tissue Eng. 2007;13:373–83.
148. Even-Ram S, Artym V, Yamada KM. Matrix control of stem cell fate. Cell. 2006;126:645–7.
149. Singh M, Haverinen HM, Dhagat P, Jabbour GE. Inkjet printing-process and its applications. Adv Mater. 2010;22:673–85.
150. Sekitani T, Noguchi Y, Zschieschang U, Klauk H, Someya T. Organic transistors manufactured using inkjet technology with subfemtoliter accuracy. Proc Natl Acad Sci USA. 2008;105:4976–80.
151. Roth Ea, Xu T, Das M, Gregory C, Hickman JJ, Boland T. Inkjet printing for high-throughput cell patterning. Biomaterials. 2004;25:3707–15.
152. Mironov V, Boland T, Trusk T, Forgacs G, Markwald RR. Organ printing: computer-aided jet-based 3D tissue engineering. Trends Biotechnol. 2003;21:157–61.
153. Morgan S, Rose F, Matcher S. Optical techniques in regenerative medicine. 2013. doi:10.1201/b15324.
154. Vielreicher M, Schürmann S, Detsch R, Schmidt Ma, Buttgereit A, Boccaccini A, Friedrich O. Taking a deep look: modern microscopy technologies to optimize the design and functionality of biocompatible scaffolds for tissue engineering in regenerative medicine. J R Soc Interface. 2013;10:20130263.
155. Wang TD, Friedland S, Sahbaie P, Soetikno R, Hsiung PL, Liu JTC, Crawford JM, Contag CH. Functional imaging of colonic mucosa with a fibered confocal microscope for real-time in vivo pathology. Clin Gastroenterol Hepatol. 2007;5:1300–5.
156. Pysz MA, Gambhir SS, Willmann JK. Molecular imaging: current status and emerging strategies. Clin Radiol. 2010;65:500–16.
157. Andreu Z, Khan MA, González-Gómez P, et al. The cyclin-dependent kinase inhibitor p27 kip1 regulates radial stem cell quiescence and neurogenesis in the adult hippocampus. Stem Cells. 2015;33:219–29.
158. Zhou Y, Liu X, Engstrom EM, Nimchuk ZL, Pruneda-Paz JL, Tarr PT, Yan A, Kay Sa, Meyerowitz EM. Control of plant stem cell function by conserved interacting transcriptional regulators. Nature. 2014;517:377–80.
159. Teo GSL, Yang Z, Carman CV, Karp JM, Lin CP. Intravital imaging of mesenchymal stem cell trafficking and association with platelets and neutrophils. Stem Cells. 2015;33:265–77.
160. Kempe M, Rudolph W. Comparative study of confocal and heterodyne microscopy for imaging through scattering media. J Opt Soc Am A. 1996;13:46–52.
161. Peng T, Xie H, Ding Y, Wang W, Li Z, Jin D, Tang Y, Ren Q, Xi P. CRAFT: multimodality confocal skin imaging for early cancer diagnosis. J Biophotonics. 2012;5:469–76.
162. Ivanov DP, Parker TL, Walker Da, Alexander C, Ashford MB, Gellert PR, Garnett MC. In vitro co-culture model of medulloblastoma and human neural stem cells for drug delivery assessment. J Biotechnol. 2015;205:1–11.

163. Tong PL, Roediger B, Kolesnikoff N, Biro M, Tay SS, Jain R, Shaw LE, Grimbaldeston Ma, Weninger W. The skin immune atlas: Three-dimensional analysis of cutaneous leukocyte subsets by in vivo microscopy. J Invest Dermatol. 2014. doi:10.1038/jid.2014.289.
164. Park Y II, Lee KT, Suh YD, Hyeon T. Upconverting nanoparticles: a versatile platform for wide-field two-photon microscopy and multi-modal in vivo imaging. Chem Soc rev. 2014. doi:10.1039/c4cs00173g.
165. Kobat D, Horton NG, Xu C. In vivo two-photon microscopy to 1.6-mm depth in mouse cortex. J Biomed Opt. 2011;16:106014.
166. Nam SY, Ricles LM, Suggs LJ, Emelianov SY. Imaging strategies for tissue engineering applications. Tissue Eng Part B Rev. 2014;21(1):88–102.
167. Fujimoto JG. Optical coherence tomography for ultrahigh resolution in vivo imaging. Nat Biotechnol. 2003;21:1361–7.
168. Yang Y, Dubois A, Qin X, Li J, El Haj A, Wang RK. Investigation of optical coherence tomography as an imaging modality in tissue engineering. Phys Med Biol. 2006;51:1649–59.
169. Lumbroso B, Huang D, Romano A, Rispoli M, Coscas G, Editors. Clinical En Face OCT Atlas, 1st ed. 2013. pp. 77–8.
170. Rompolas P, Deschene ER, Zito G, Gonzalez DG, Saotome I, Haberman AM, Greco V. Live imaging of stem cell and progeny behaviour in physiological hair-follicle regeneration. Nat. 2012;487:496–9.
171. Assayag O, Grieve K, Devaux B, Harms F, Pallud J, Chretien F, Boccara C, Varlet P. Imaging of non-tumorous and tumorous human brain tissues with full-field optical coherence tomography. NeuroImage Clin. 2013;2:549–57.
172. Subhash HM, Connolly E, Murphy M, Barron V, Leahy M. Photothermal optical coherence tomography for depth-resolved imaging of mesenchymal stem cells via single wall carbon nanotubes. Appl. San Fransisco: SPIE. Nanoscale Imaging, Sensing, Actuation Biomed; 2014. p 89540C.
173. Wang LV. Prospects of photoacoustic tomography. Med Phys. 2008;35:5758.
174. Emelianov SY, Li P-C, O'Donnell M. Photoacoustics for molecular imaging and therapy. Phys Today. 2009;62:34–9.
175. Margolis DJa, Hoffman JM, Herfkens RJ, Jeffrey RB, Quon A, Gambhir SS. Molecular imaging techniques in body imaging. Radiology. 2007;245:333–56.
176. Ballyns JJ, Bonassar LJ. Image-guided tissue engineering. J Cell Mol Med. 2009;13:1428–36.
177. Hillman EMC, Burgess SA. Sub-millimeter resolution 3D optical imaging of living tissue using laminar optical tomography. Laser Photonics. 2009;179:159–79.
178. Burgess SA, Ratner D, Chen BR, Hillman EMC. Fiber-optic and articulating arm implementations of laminar optical tomography for clinical applications. Biomed Opt Express. 2010;1:780–90.
179. Ozturk MS, Lee VK, Dai G, Intes X, Zhao L. Laminar optical tomography applied to reporter genes imaging in engineered tissue constructs. Northeast Bioeng. Conf. Syracuse, NY, 2013. pp 129–130.
180. Björn S, Englmeier K-H, Ntziachristos V, Schulz R. Reconstruction of fluorescence distribution hidden in biological tissue using mesoscopic epifluorescence tomography. J Biomed Opt. 2011;16:046005.
181. Cherry SR, Shao Y, Silverman RW, et al. MicroPET: a high resolution PET scanner for imaging small animals. IEEE Trans Nucl Sci. 1997;44:1161–6.
182. Wong MD, Dazai J, Walls JR, Gale NW, Henkelman RM. Design and implementation of a custom built optical projection tomography system. PLoS ONE. 2013. doi:10.1371/journal.pone.0073491.

Bioprinting with Live Cells

S. Burce Ozler, Can Kucukgul and Bahattin Koc

1 Introduction

The loss or failure of an organ or tissue is one of the most devastating, and costly problems in health care. The current treatment methods for organ/tissue loss or failure include transplantation of organs, surgical reconstruction, use of mechanical devices, or supplementation of metabolic products. Due to the growing need for organ transplantation and a lack of donor organs, tissue or organ engineering has progressed as a multidisciplinary field combining life sciences and engineering principles to restore, maintain, or improve function of tissues or organs [1, 2].

Traditionally, tissue engineering strategies are based on the cell seeding into synthetic, biological or composite scaffolds providing a suitable environment for cell attachment, proliferation and differentiation. A scaffold is highly porous complex structure providing an interconnected network that is designed to act as an artificial extracellular matrix (ECM) until the cells form their own ECM. In scaffold-based tissue engineering, three steps must be followed including finding a source of precursor cells from the patient, seeding these cells in vitro into the desired places of scaffold and surgically implanting the scaffold into the patient [3]. Scaffolds have been used to fabricate various tissue grafts including skin, cartilage, bone, blood vessels, bladder and myocardium [4–12]. Bioengineered tissue scaffolds attempt to mimic both the external shape and internal architecture of the replaced tissues. The modeling of scaffolds has a great impact on the growth and proliferation of cells and a spatially and temporally controlled scaffold design could improve cell growth and differentiation [13]. Although many different scaffold manufacturing techniques such as salt-leaching, porogen melting, gas foaming, electrospinning,

B. Koc (✉) · S. B. Ozler · C. Kucukgul
Department of Manufacturing and Industrial Engineering,
Sabanci University, Istanbul, Turkey
e-mail: bahattinkoc@sabanciuniv.edu

K. Turksen (ed.), *Bioprinting in Regenerative Medicine,* Stem Cell Biology and Regenerative Medicine, DOI 10.1007/978-3-319-21386-6_3

fiber deposition, molding and freeze-drying have been investigated in the past, it is challenging to control pore size, porosity, pore interconnectivity and external geometries of scaffolds. In recent years, various additive manufacturing based methods such as bioplotting, bioprinting, ink-jet printing and stereolithography have been used for biomanufacturing of complex three-dimensional (3D) tissue scaffolds to overcome the limitations of conventional tissue engineering methods [14]. These additive manufacturing based techniques allow to fabricate scaffolds layer-by-layer with controlled external and internal geometries based on computer-aided models of targeted tissues [15]. Several researchers have investigated designing functionally gradient porous scaffolds with controllable variational pore size and heterogeneous porous architecture [16, 17].

In scaffold-based tissue engineering, different biomaterials are used for scaffold fabrication such as porous materials composed of biodegradable polymers (polylactic acid, polyglycolic acid, hyaluronic acid and several copolymers), hydroxyapatite or calcium phosphate-based materials and soft materials like collagens, fibrin, and various hydrogels. Although there is a plenty of choice for scaffold materials, an ideal biomaterial for scaffold fabrication should be nontoxic, nonimmunogenic, capable of maintaining mechanical integrity for tissue growth and differentiation with controlled degradation [3].

After implantation of a scaffold, it should degrade in a controlled manner and the seeded cells should proliferate and migrate into scaffold to replace the scaffold biomaterial. Newly-formed extracellular matrix (ECM) fills the places which were previously occupied by the biomaterial of scaffolds. However, there are some drawbacks to create tissues with biodegradable scaffolds. Mostly, oxygen/nutrient delivery and removal of metabolic waste are insufficient through the micro-channels of a scaffold. Additionally, biodegradation of the scaffold induces inflammation. Even though the biomaterials used may not be directly toxic, they can be metabolized to toxic byproducts [18].

Because of the above mentioned drawbacks, the recent tissue engineering studies tend towards ‘scaffold-free' techniques. During the embryonic maturation, tissues and organs are formed without the need for any scaffolds [19, 20]. The self-assembly and self-organizing capabilities of cells and tissues are main driver for the field of scaffold-free tissue engineering. Self-assembly based tissue engineering aims to produce fully biological tissues with specific compositions and shapes having the ability to grow their own ECM, and therefore to reduce the immunogenic reactions and other unpredicted complications based on the use of scaffolds [21].

One way of implementing the self-assembly approach is the cell sheet technology, which has been applied clinically for the repair of skin, cornea, blood vessels, and cardiomyocyte patches to repair partial heart infarcts [18, 22, 23]. Another self-assembly-based approach is founded on the recognition that ‘nature knows best'. This approach relies on the principle that cell aggregates can be used as building blocks, since they have the intrinsic capacity to fuse together, known as tissue fluidity and self-assemble through morphogenetic processes if they are deposited in close spatial organization [24–26]. The engineering of 3D living structures supported by the self-assembly and self-organizing capabilities of cells is commonly termed 'bioprinting'. Bioprinting is an extension of tissue engineering, where the cells are delivered through the application of additive manufacturing techniques [27, 28]

This chapter focuses on scaffold-free tissue engineering and its adaptation to the technology of three dimensional bioprinting. Further, the importance as well as the challenges for 3D bioprinting using stem cells will be discussed in this chapter.

2 Bioprinting with Live Cells

2.1 2D Patterning and Cell-Sheet Technology

Placing cells into special patterns using the laser light has been one of the first methods used for 2D cell patterning [14]. These laser-based techniques utilize transparent ribbons on which one side is coated with cells that are either adhered to a biological polymer through initial cellular attachment or uniformly suspended in a thin layer of liquid or a hydrogel. A pulsed laser beam is transmitted through the ribbon and is used to push cells from the ribbon to the receiving substrate which is coated with hydrogels.

While laser based approaches enable to pattern living cells on a substrate [29] and to layer multiple cell types [30], laser-based techniques have been also explored for positioning of cells in microarrays [31]. The resolution of laser-assisted bioprinting is affected by different factors such as the laser fluence, the wettability of the substrate, and the thickness and the viscosity of the biological layer [32]. Guillotion and his group studied the effect of the viscosity of the bioink, laser energy, and laser printing speed on the resolution of cell printing [33] as shown in Fig. 1.

By using this method, various cell types including human osteosarcama, rat cardiac cells and human umbilical vein endothelial cells (HUVEC) could be printed with micrometer accuracy on Matrigel as the absorptive layer [30, 34, 35]. More recently, the biological laser printing was used to print sodium alginate, nano-sized hydroxyapatite (HA) and human endothelial cells [36]. However, most of these methods are limited to two-dimensional patterning and it is difficult to fabricate three-dimensional tissue constructs because of process-induced cell injury. The thermal stress and ultraviolet radiation caused by laser printing could also affect the cell viability.

Similar to 2D patterning, cell sheet technology is another scaffold-free method for construction of 2D and 3D engineered tissues. In this method, cells can be removed from a culture dish as a relatively stable confluent monolayer sheet without destroying cell-cell contacts. In order to build a substantial 3D tissue volume, many sheets need to be culminated in high amount of cells which requires vascularization for cell viability [15].

L'Heureux and his group produced a tissue engineered blood vessel using a cell-sheet approach based on cultured human cells. The developed vessel contained all three histological layers such as the endothelium, the media and the adventitia. In

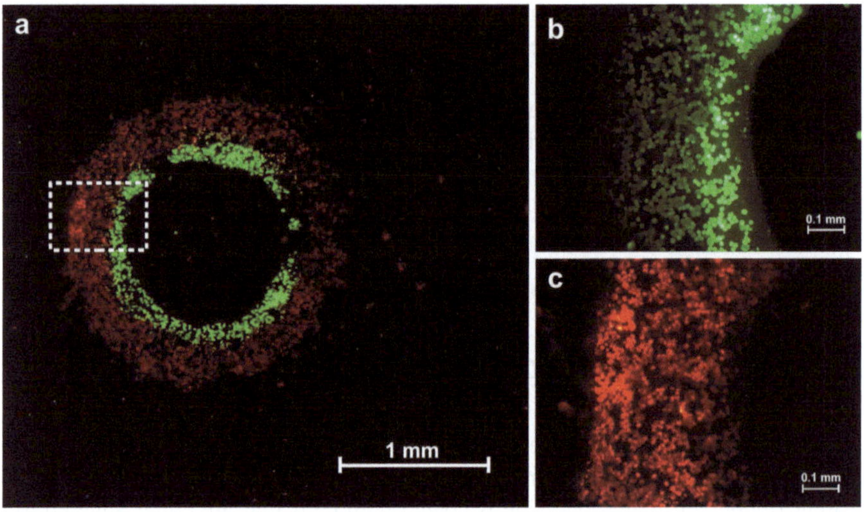

Fig. 1 Laser-assisted sequential two color cell printing in 2D. **a** The two cell suspensions (6·107 cells/ml in DMEM, supplemented with 1% (w/v) alginate were then printed according to a pattern of concentric circles). **b** *Green* Calcein stained cells within the region of interest as defined in 1**a** (*dashed rectangle*). **c** *Red* fluorescent DiI-LDL stained cells within the same region of interest. (Adapted from [33])

this self-assembly approach, smooth-muscle cells and fibroblasts were cultured in medium containing serum and ascorbic acid and produced their own extracellular matrix (ECM). The smooth-muscle cell sheet was placed around a tubular mandrel to produce the media of the vessel. A similar fibroblast cell sheet was wrapped around the media to provide the adventitia after 8 weeks of maturation. Finally, the tubular support was removed and endothelial cells were seeded in the lumen to form the endothelium. The tissue engineered blood vessel has burst strength of over 2500 mm Hg which is significantly higher than that of human saphenous veins (1680 ± 307 mm Hg) [23]. Sheet-based tissue engineering has been used by the same group to produce tissue engineered blood vessel (TEBV) suitable for autologous small diameter arterial revascularization in adult patients. Fibroblasts were cultured in conditions promoting extracellular matrix (ECM) deposition to produce a cohesive sheet that can be detached from the culture flask. This approach also eliminates the use of smooth-muscle cells, whose early senescence is related with decreased burst pressures in human models. The decellularized internal membrane (IM) and living adventitia were assembled by wrapping fibroblast sheets around a temporary Teflon coated stainless steel support tube. After weeks-long maturation and dehydration to form an acellular substrate, the steel tube was removed and endothelial cells were seeded in the lumen of living TEBV. The transplantation of these blood vessels into dogs demonstrated good handling, suturability by the use of conventional surgical techniques. Ultimately, this is an effective approach to produce a completely biological and clinically applicable TEBV in spite of its relatively long production time (≈28 weeks) which clearly prevents its urgent clinical use [22].

Okano and colleagues have engineered a long-lasting cardiac tissue based on a similar self-assembled sheet based approach. In their method, culture dishes are first coated with a temperature-responsive polymer, poly (N-isopropylacrylamide) (PIPAAm). The surface is relatively hydrophobic at 37 °C allowing cells to attach and proliferate, while cooling below 32 °C (typically 20 °C for 30 min) makes the surface hydrophilic and causes the cells to detach without the use of enzyme digestion reagent. When grafted PIPAAm layer thickness is between 15 and 20 nm, temperature-dependent cell adhesion and detachment can be observed. Once the cells spread and confluent on the surface, they can be spontaneously detached as a contiguous cell sheet by reducing the temperature. This process does not disrupt the cell-cell junctions because no enzymes like trypsin are required. Additionally, basal surface extracellular matrix (ECM) proteins such as fibronectin are preserved after detachment which enables easy attachment of cell sheets to host tissues and even wound sites with minimal cell loss. In order to obtain tissue constructs with characteristic physiological cellular functions in vitro, heterotypic cell-cell interactions are inevitable. As shown in Fig. 2, it is possible to modify the above-mentioned technique in order to develop patterned cell sheets using two or more kinds of cell source. Domains on petri dishes were grafted by using area-selective electron beam polymerization of PIPAAm. After cells were cultured on the patterned grafted surfaces at 37 °C, the temperature was decreased to 20 °C. Cells on the PIPAAm surface are detached where other cell types were seeded subsequently by increasing the temperature to 37 °C. Therefore, two cell types can be co-cultured in desired places which improve cellular functions [18, 37, 38]. Three-dimensional myocardial tubes

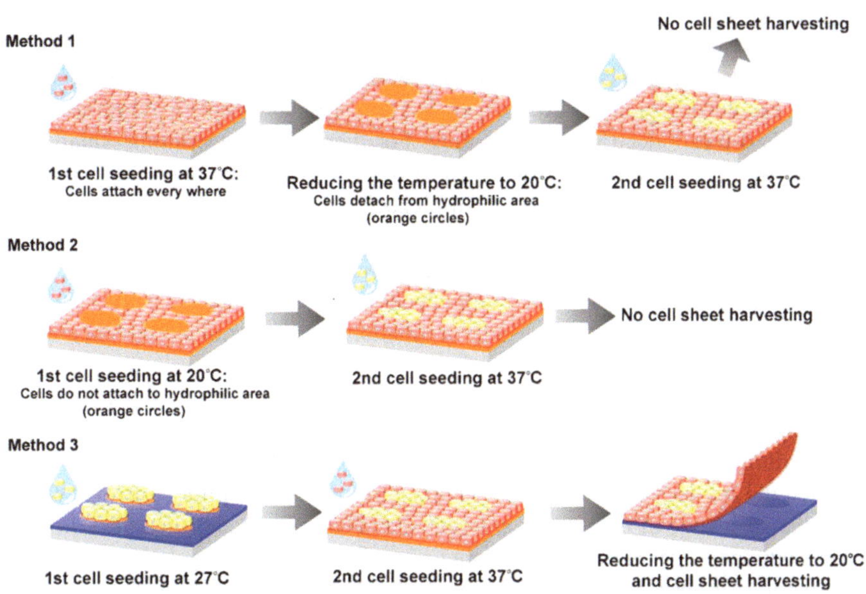

Fig. 2 Schematic diagram of methods of cell seeding for patterned co-culture on PIPAAm-grafted surfaces. (Adapted from[39])

were fabricated using neonatal rat cardiomyocyte sheets cultured on temperature-responsive culture dishes [39, 40]. Due to the functional gap junction formation, electrical coupling of cardiomyocyte sheets was obtained quickly and the construct was implanted [41]. Four weeks after the implantation, the myocardial tubes were integrated with the host tissues showing spontaneous and synchronized pulsation [37, 42]. Using this versatile method, functional and transplantable tissue sheets are produced from different cell types including epidermal keratinocytes [43], kidney epithelial cells [44] and periodontal ligaments [45, 46].

Two-dimensional cell patterning or cell-sheet based approaches have been successful tissue engineering approaches. However, the engineered constructs fabricated with these methods are limited to 2D cell patterns or simple shapes because of the flat and uncontrolled shape of the cell sheets. In addition, many sheets also need to be culminated in high amount of cells which requires pre-vascularization of the sheets for 3D tissue constructs. Therefore, several bioprinting approaches have been developed for fabricating 3D tissue constructs with live cells. Two major approaches, ink-jet based and extrusion based 3D bioprinting methods will be explained in details below.

2.2 *Inkjet-Based Bioprinting*

In inkjet-based bioprinting, a bioink made of cells and biomaterials are printed in the form of droplets through cartridges onto a substrate. There are two types of inkjet printing including continuous inkjet printing (CIJ) and drop-on-demand inkjet printing (DOD). In CIJ mode, a jet is obtained by forcing the liquid through an orifice under an external pressure and it breaks up into a stream of droplets. In DOD mode, a pressure pulse is applied into the fluid which generates drops only when needed. For ink-jet printing of cells, there are two most commonly used approaches: thermal and piezo-electric inkjet printing. For thermal inkjet printing, small volumes of the printing fluid are vaporized by a micro-heater to create the pulse that expels droplets from the print head. In piezoelectric inkjet printing, a direct mechanical pulse is applied to the fluid in the nozzle by a piezoelectric actuator, which causes a shock wave that forces the bioink through the nozzle [47]. However, there have been only a few examples of cell deposition by piezoelectric ink-jet printing due to the electrically conducting ink formulations and contamination concerns on ink recycling [24].

Inkjet-based bioprinting (Fig. 3) enables to deposit different cell types in precise orientations relative to the print surface and to each other at micrometer resolution by controlling the ejection nozzles and timing of spray [48]. Wilson and Boland first adapted the ink-jet printers for the manufacture of cell and protein arrays, which have the advantage of being fully automated and computer controlled [49]. In their next study, cell aggregates were printed onto thermosensitive gels layer-by-layer in order to demonstrate the fusion between the closely-placed cell aggregates [50]. The same group deposited CHO cells and rat embryonic motoneurons as an 'ink'

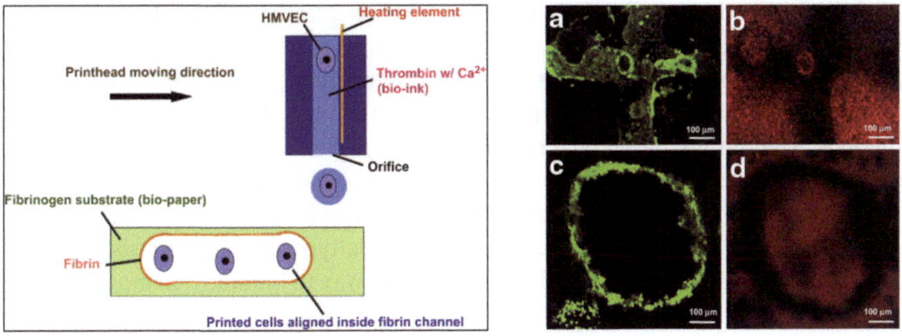

Fig. 3 Schematic diagram of inkjet bioprinting methods of cells with fibrin channel scaffold printed microvasculature. **a** Printed perpendicular microvasculature cultured for 14 days. **b** Integrity of the printed structure stained using Texas *Red* conjugated dextran molecules of 3000 MW. **c** Printed *ring* shaped microvasculature cultured for 21 days. **d** Integrity of printed structure cultured for 21 days

onto several 'bio-papers' made from soy agar and collagen gel. They demonstrated also that the mammalian cells can be effectively delivered by a modified thermal inkjet printer onto biological substrates and that they retain their ability to function [51]. Cui and Boland used also thermal inkjet printing to produce cell containing fibrin channels by printing human microvascular endothelial cells onto thin layers of fibrinogen as shown in Fig. 3. During the incubation period, the cells proliferated and formed branched tubular structures mimicking simple vasculature [52].

Inkjet bioprinting has been progressed to fabricate 3D biological structures by the use of readily crosslinked hydrogels such as alginate. In published studies, cells have been mixed with alginate solutions and crosslinked with calcium chloride to create cell encapsulating hydrogels having defined 3D structures [53–55]. In a more recent study, alginate has been used as a constituent of bioink and it was mixed with NIH 3T3 mouse fibroblasts cell suspension in order to fabricate zigzag cellular tubes with an overhang structure using a platform-assisted 3D inkjet bioprinting system [56].

Although inkjet bioprinting has been one of the most commonly used method in printing living cells and biomaterials, cell aggregation, sedimentation and cell-damage because of the high shear stresses are common drawbacks of this method. Cell aggregation and sedimentation may be prevented by frequent stirring of the cell mixture, which can result in reduced cell viability if the cells are sensitive to the shear forces [57]. Another problem limiting the inkjet bioprinting is the clogging of the nozzle orifice. Low viscosity surfactants can be added to the ink which can cause additional challenges such as cell damage [58].

Recently, two research groups successfully address the sedimentation and cell aggregation problem during the inkjet bioprinting. Chahal and coworkers used a surfactant (Ficoll PM400) create neutrally buoyant MCF-7 breast cancer cell suspensions, which were ejected using a piezoelectric drop-on-demand inkjet printing system. They demonstrated that Ficoll PM400 did not have adverse effects on cell viability. Moreover, neutrally buoyant suspension greatly increased the reproducibility of consistent cell counts, and eliminated nozzle clogging. [59].

Ferris et al. used two different commercially available drop-on-demand printing systems in order to reproducibly print several different cell types over long printing periods. The bio-ink based on a novel microgel suspension in a surfactant-containing tissue culture medium can prevent the settling and aggregation of cells, while meeting the stringent fluid property requirements needed for many-nozzle commercial inkjet print heads. They could print two cells types simultaneously from two different inkjet print heads, which is a innovative way to biofabricate more complex multi-cellular structures [60].

2.3 Self-Assembly Based Bioprinting

The autonomous organization of components from an initial state into a final pattern or structure without external intervention is called self-assembly. The aim of the self-assembly-based bioprinting is the use of the inherent organizational capacity of cells into tissues and eventually organs by mimicking natural morphogenesis. The best examples of tissue self-organization and self-assembly are in the field of developmental biology and scaffold-free biomimetic approach has deep roots in developmental biology [20]. Malcolm Steinberg published papers, in which he formulated fundamental thermodynamic rules determining tissue self-assembly and developed differential adhesion hypothesis (DAH) explaining the fluidic nature of cell sorting and tissue self-assembly [61–63]. Therefore, the novel scaffold-free biomimetic tissue engineering paradigm relies on the principle that in vitro tissue assembly from single cells or tissue aggregates is feasible.

Based on the self-assembly principle, it is possible to fabricate reliable and reproducible 3D tissue constructs having defined topology and functionality in vitro when combined with bioprinting techniques. The disadvantage of this method is that the development of the natural ECM is time consuming and in vitro self-assembly may vary with fully physiological conditions. The bioprinting of 3D tissue constructs is achieved via a three-phase process: (1) preprocessing or bio-ink preparation; (2) processing, i.e. the actual automated deliver/printing of the bio-ink particles into the bio-paper by the bioprinter; and (3) postprocessing, i.e. the maturation/incubation of the printed construct in the bioreactor [19]. Self-assembly occurs in an *in vivo* like, fully controllable cell environment (bioreactor) by the differentiation of cells at the right time, in the right place and into the right phenotype and eventually the assembly of them to form functional tissues. Based on this approach, a perfusion reactor is used for the maturation of a bioprinted macrovascular network in order to obtain the required mechanical properties. Microvascular units consisting of cylindrical or spherical multicellular aggregates were fabricated by the parenchymal and endothelial cells. Afterwards, microvascular units were located in the macrovascular network for the perfusion supporting self-assembly and the connection to the existing network. Multicellular spherical and cylindrical aggregates have been constructed by using 3D printing methods, which enable to achieve flexibility in tube diameter and wall thickness and to form branched tubular structures. However, the printed cell aggregates should be perfectly supported by hydrogels for 3D printing [21]. Forgacs and his group employed this novel technology to print cellular topologically defined structures of various shapes. Cardiac constructs were built using embryonic cardiac and endothelial cells and their postprinting self-assembly resulted in synchronously beating solid tissue blocks, where the endothelial cells were organized into vessel-like conduits [64]. In their more recent study, the same group utilized the self-assembly approach in order to bioprint small-diameter, multilayered, tubular vascular and nerve grafts using bio-ink composed of aortic smooth muscle cells (HASMC), human aortic endothelial cells (HAEC), human dermal fibroblasts (HDFb) and bone marrow stem cells (BMSC), respectively as shown in Fig. 4 [25]. Similarly, in another study, self-assembled cell-based microtissue blocks were used to generate small diameter tissue-engineered living blood vessels (TEBV). Microtissues composed of human-artery-derived fibroblasts (HAFs) and endothelial celss (HUVECs) were cultured for 7 and 14 day under pulsatile flow/mechanical stimulation in a designed bioreactor or static culture conditions with a diameter of 3 mm and a wall thickness of 1 mm. Self-assembled microtissues

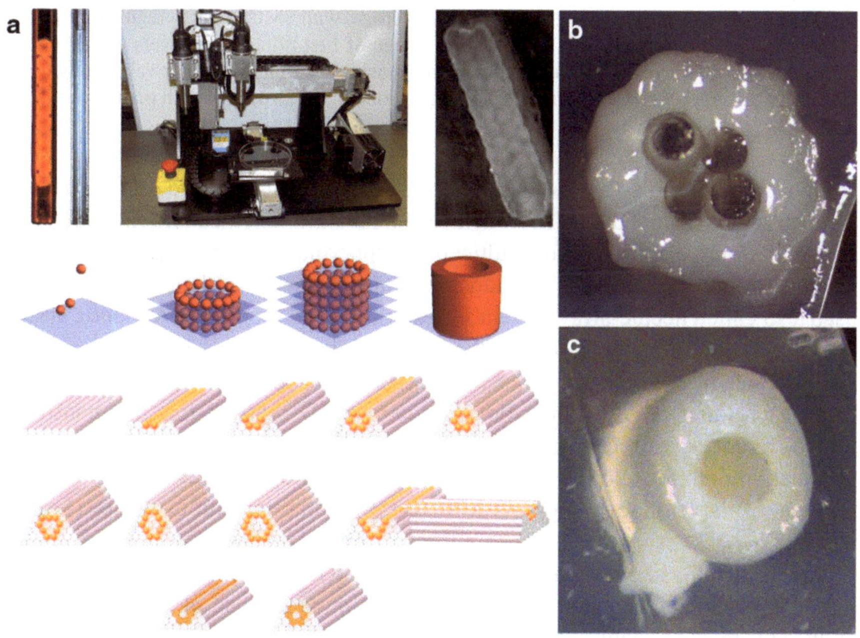

Fig. 4 **a** Organovo bioprinter with cell and hydrogel printing heads and schematics to print tubular structures with cellular spheroids: layer-by-layer deposition of spheroids into the hydrogel. **b** Cross-section of a vascular graft printed with four central rods 12 h post-printing. All cellular cylinders have fused to form a continuous conduit. **c** A vascular construct (ID =600 μm) at 14 days post-perfusion. (Adapted from [25])

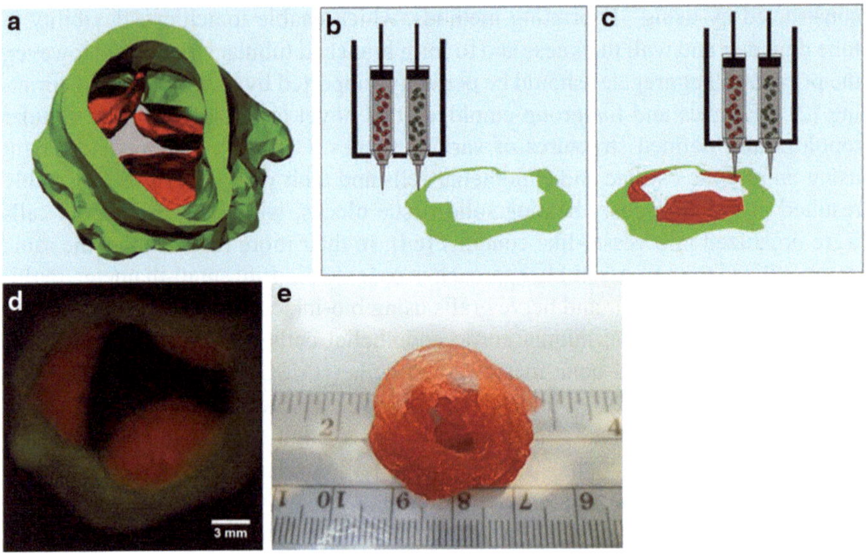

Fig. 5 Bioprinting of aortic valve conduit. **a** Aortic valve model reconstructed from micro-CT images. The root and leaflet regions were identified with intensity thresholds and rendered separately into 3D geometries into STL format (*green* color indicates valve root and *red* color indicates valve leaflets); **b**, **c** schematic illustration of the bioprinting process with dual cell types and dual syringes; **b** root region of first layer generated by hydrogel with SMC; **c** leaflet region of first layer generated by hydrogel with VIC; **d** fluorescent image of first printed two layers of aortic valve conduit; SMC for valve root were labeled by cell tracker *green* and VIC for valve leaflet were labeled by cell tracker *red*. **e** as-printed aortic valve conduit. (Reproduced with permission [74]).

composed of fibroblasts showed accelerated ECM formation and a layered tissue formation was obtained only in flow/mechanical stimulation conditions [65].

An alternative approach for multicellular spheroid assembly technique for biofabrication was developed by Nakayama and his group. In their technique, they used a so called needle-array system instead of using a bioprinter. A robotic system was developed in order to skewer the multicellular spheroids into medical-grade stainless needles, which served as temporal fixators until multicellular spheroids fused each other. They could fabricate complex 3D scaffold-free cell constructs with different types of cells including chondrocyte, hepatocyte, cardiomyocyte and vascular smooth muscle cell. One of the advantages of this technique is the easy removal of the temporary supports without contamination with exogenous materials [3].

Apart from the above mentioned applications, a new method is presented to rapidly self-assemble cells into 3D tissue rings without the use of additional equipment. This method enables fabrication of engineered tissue constructs entirely from cells by seeding cells into custom made annular agarose wells with 2, 4 or 6 mm inside diameters. Different cell types including rat aortic smooth muscle cells and human smooth muscle cells are used with varying seeding conditions and culture length to form tissue rings. The strength and modulus of tissue rings increased with ring size and decreased with culture duration. Rat smooth muscle cell rings with an inner diameter of 2 mm are cohesive enough for handling after 8 days incubation and they yield at 169 kPa ultimate tensile strength. Furthermore, it is also possible to fabricate tissue tubes by transferring the rings onto silicone tubes, sliding them into contact with one another and incubating them for an additional 7 days. Although these rings are not as strong as ring segments of native blood vessels or TEBV fabricated from cell sheets for 2–3 months, the presented method allows developing 3D tissue constructs from aggregated cells within an experimentally useful time frame (1–2 weeks) [66]. Likewise created smooth muscle cell tissue rings and rings fabricated from cells seeded in fibrin or collagen gels are compared based on their relative strength and utility for tissue engineering. All tissue rings were cultured for 7 days in supplemented growth medium which includes ε-amino caproic acid, ascorbic acid, and insulin-transferrin-selenium. Ultimate tensile strength and stiffness values of tissue rings were two-fold higher than fibrin gel and collagen gel rings. Tissue rings cultured in supplemented growth medium exhibit a three-fold increase in tensile strength and stiffness in comparison to the tissue rings cultured in standard growth medium [67].

The approach of using microtissues as building blocks to form larger structures is further used by other research groups in order to investigate the reassemble capacity of cell aggregates. After a preculture period for 7 days of HUVEC spheroids, they were mixed with NHF cells and were able to reassemble and form microtissues with the NHF cells on the inside and coated with HUVEC on the outside [68]. Additionally, the kinetics of the cellular self-assembly also differs from one cell type to the other. While H35 cells formed relatively stable rod structures inside the recesses of micromolded agarose gels, NHF cells reassembled quickly the initial rod structures to a final spheroid structure [69].

2.4 Extrusion-Based Bioprinting

Bioprinting methods based on extrusion of cell or cell-laden biomaterials use self-assembly cells to construct 3D biological constructs. The main principle of extrusion-based bioprinting techniques is to force continuous filaments of materials including hydrogels, biocompatible copolymers and living cells through a nozzle with a help of a computer to construct a 3D structure [27]. Extrusion-based printers usually have a temperature-controlled material handling and dispensing system and stage with the movement capability along the x, y and z axes. The printers are directed by the CAD-CAM software and continuous filaments are deposited in two dimensions layer-by-layer to from 3D tissue constructs. The stage or the extrusion head is moved along the z axis, and the printed layers serve as a base and support for the next layer. Pneumatic or mechanical (piston or screw) are the most common techniques to print biological materials for 3D bioprinting applications [32]. Additionally, novel multi-nozzle biopolymer deposition systems were developed for freeform fabrication of biopolymer-based tissue scaffolds and cell-embedded tissue constructs [70, 71].

An extrusion-based printer was used to deposit living cells by Williams and co-workers. Instead of using a thermally crosslinked biomaterial, which can flow at room temperature, but crosslink into a stable material at body temperature, they used Pluronic F-127 and type I collagen to encapsulate human fibroblasts and bovine aortic endothelial cells (BAECs) separately. These materials flow at physiologically suitable temperatures (35–40°), but crosslink at room temperature. They demonstrated the availability of CAD/CAM technology to fabricate anatomically correct shaped constructs and also examined several environmental factors with respect to the viability of the extruded cells [72, 73]. Recently, different research groups used the similar extrusion systems in order to fabricate anatomically accurate and mechanically heterogeneous aortic valves as shown in Fig. 5 [74, 75].

Several groups used high resolution extrusion systems to print different type of cells encapsulated in various hydrogels. For instance, Chang et al. printed HepG2 cell encapsulated sodium alginate through a pneumatically powered nozzle and examined the process parameters, the dispensing pressure and the nozzle diameter, regarding to the cell viability and recovery [76]. In another study, alginate hydrogel was used with calcium sulfate as a crosslinking agent to fabricate pre-seeded implants of arbitrary geometries and the printed constructs showed high viabilities [77]. Although the cell viability after printing is important, it is also important that the cells perform their essential functions in the tissue constructs.

Extrusion-based printing allows the construction of organized structures within a realistic time frame, and hence it is the most promising bioprinting technology. The main advantage of extrusion-based bioprinting is the ability to print very high cell densities. Some groups developed 3D bioprinters in order to use multicellular spheroids or cylinders as bioink to create 3D tissue constructs [19, 21, 25, 78–80]. However, preparing bioink requires time-consuming manual operation and makes totally automated and computer-controlled 3D bioprinting impossible in earlier studies. Therefore, our group focused on the development of a continuous bioprinting approach in order to extrude cylindrical multicellular aggregates using an extrusion-based bioprinter, which is an automated, flexible platform designed to fabricate 3D tissue engineered cell constructs. In order to bioprint anatomically correct tissue constructs directly from medical images, the targeted tissue or organ must be biomodeled. In the following section, the details of modeling and developing path planning for automated direct cell bioprinting will be explained.

2.4.1 Biomodeling

In order to obtain an anatomically accurate tissue constructs, several imaging methods for data acquisition of tissue organ such computed tomography (CT) and magnetic resonance imaging (MRI) could be used. The obtained medical images are then transferred to a special segmentation software, where the images are represented with stack of numerous planar scan captures (Fig. 6a). The segmented 3D

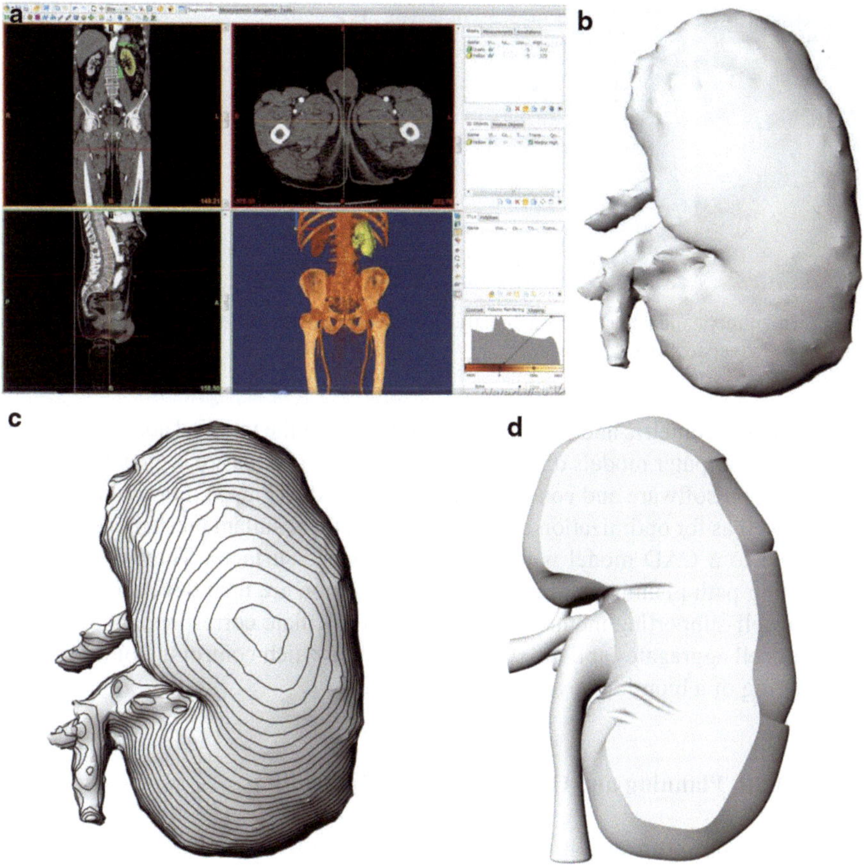

Fig. 6 Biomodeling of the bioprinted tissue/organ **a** segmentation of the targeted tissue/organ **b** mesh model of segmented model **c** slicing of CAD model **d** a layer of CAD model

surface geometry is transformed to a CAD model which is a mesh model of the object (Fig. 6b). In order to generate bioprinting path planning as well as the topology optimization for bioprinting processes, the resultant mesh models need to be represented by smooth parametric surfaces. The mesh model is then sliced with consecutive planar cross-sections, resulting in closed contour curves for each thin layer slice [] (Fig. 6c). Those contour curves are basically the surface boundaries of tissue constructs. Obtained contour curves need to smoothed by B-spline curve fitting from their control points, in order to generate smooth parametric surfaces and finer surface geometry (Fig. 6d). The resultant CAD model is then ready to be used for path planning and topology optimization purposes for biomimetic 3D bioprinting.

Recently, novel computer-aided algorithms and strategies are developed to model and 3D bioprint a scaffold-free human aortic tissue construct biomimetically by our group. Medical images obtained from magnetic resonance imaging (MRI) are used

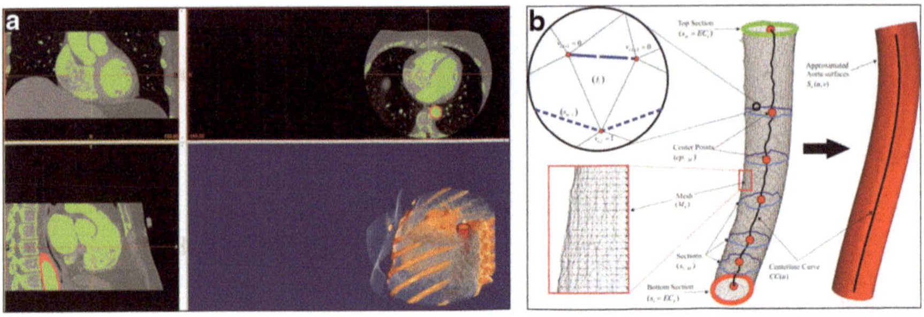

Fig. 7 Biomimetic modeling of aorta directly from medical images **a** segmentation from medical images **b** Conversion to a CAD model

to obtain the geometric and topological information of the targeted aorta. In order to obtain 3D computer models of the aortic tissue, MRI images are segmented using a segmentation software and converted into CAD model (Fig. 7). For tool path planning as well as for optimization of 3D bioprinting, the resultant mesh model of aorta converted to a CAD model with smooth parametric surfaces. Three-dimensional bioprinting path planning and parameter optimization are then developed. The developed self-supporting methodology is used to calculate corresponding tool paths for both cell aggregates and the support structures, which control the bioprinter for 3D printing of a biomimetic aortic construct [82, 83].

2.4.2 Path Planning and Optimization for Bioprinting

In order to bioprint an anatomically correct tissue constructs with live cells layer-by-layer, the generated computer model of this construct needs to be sliced by planar cross-sections, resulting in closed contour curves for each layer. Then those layers need to be filled by appropriate types of cellular aggregates with supportive hydrogel walls surrounding them for keeping the biomimetic form. In our recent work, multicellular cell aggregates are 3D bioprinted based on computer-aided continuous and, interconnected tool-path planning methodologies. Continuous bioprinting enables to design and 3D bioprint extruded multicellular aggregates according to the computer model of the targeted tissue. The Zig-zag and Contour Offsetting pattern tool-path methodologies are developed to 3D bioprint different shaped structures with multiple layers. A CAD software package was used for developing algorithms for continuous and connected bioprinting path plans. In order to keep the 3D forms of printed structures during the maturation period, a biocompatible and bio-inert agarose-based hydrogel was used as a support material [82, 83]. The developed bioprinting process starts from the bottom layer and follows the generated path plan for each particular layer consecutively through the top layer. At a layer, support material enclosing the cellular aggregates are printed first, and then cellular aggregates are deposited to fill the respective contour areas.

In Fig. 8, schematic view of pat planning strategies are showed for a layer. Figure 8a-b shows Zig-zag whereas Fig. 8c-d shows Contour Offsetting path planning.

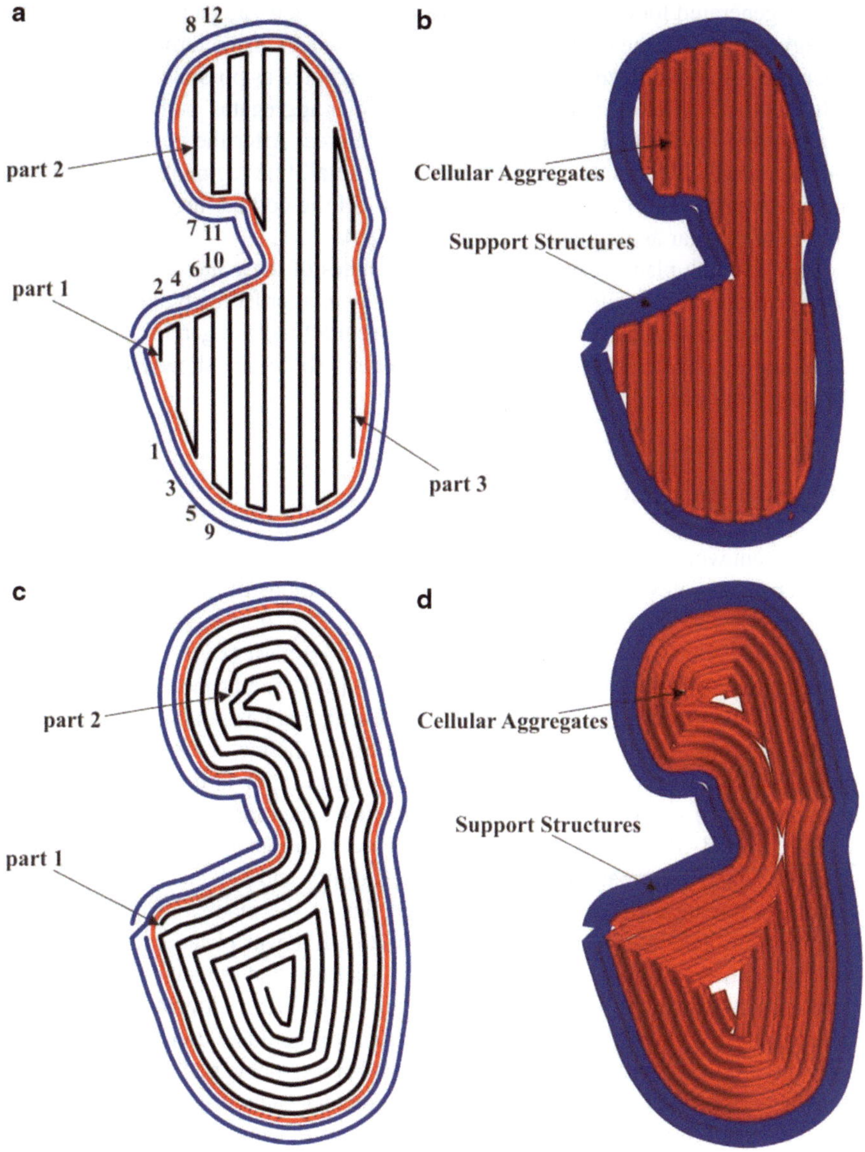

Fig. 8 Path planning for bioprinting **a–b** Zig-zag pattern **c–d** Contour offsetting path planning

In both Zig-zag and Contour Offsetting patterns, outer supportive hydrogel walls' tool path are generated with offsetting the contour curve with a deposited cellular or hydrogel extrusion diameter. Placement of support structures before the cellular aggregates provides necessary conditions for cell fusion and structure conservation.

In Zig-zag pattern path planning strategy, contour curves are crossed with parallel consecutive lines, each separated by extrusion diameter, and an intersection point is generated for each cross as shown in Fig. 8a. Zig-zag pattern path planning strategy aims to fill the contour area by following a zig-zag patterned path way between the generated intersection points as uninterrupted as possible.

In Contour Offsetting path plan strategy, cellular aggregates' path plan is formed by offsetting the contour curve to fill the entire area. After the necessary amount of offset curves is generated, they are joined by small line segments to enable continuous bioprinting. That strategy also aims to fill the contour area with minimum number of cellular aggregate as shown in Fig. 8c).

After the path plan is calculated, the coordinates of these movements are transferred to bioprinter to guide the deposition path plan in order to obtain anatomically correct tissue constructs. In Fig. 8b and Fig. 8d, the finalized generated path plans of cellular aggregates (red) and support structures (blue) are shown for both path planning strategies.

2.4.3 Continuous Cell Printing

In our recent work, a novel bioprinting method is used for precise deposition of multicellular aggregates composed of different combinations of mouse aortic smooth muscle cells (MOVAS), NIH 3T3 mouse fibroblast cells, human umbilical vein endothelial cells (HUVEC), and human dermal fibroblast (HDF) cells according to computer-generated paths as shown in Fig. 9 [82]. The proposed methodology increases the contact of cylindrical multicellular aggregates in adjacent bioprinted layers, facilitates the fusion of cells and accelerates the maturation process. More significantly, this procedure reduces the human intervention at forming of cylindrical multicellular aggregates and therefore, increases the reproducibility.

The printed 3D multicellular structures are examined for their mechanical strength, shape deformation with time, cell viability and cell fusion. The printed constructs having different shapes deformed during the incubation period (up to 10-days) and generally a shrinking between 20–38% was observed. After 4 or 7 days incubation, the support structures of well-defined and random-shaped printed structures composed of MOVAS, HUVEC and NIH 3T3 multicellular aggregates were manually removed and the fused cell structures could be transferred with forceps into a falcon filled with PBS (Fig. 10a). It is remarkable that the stripe shaped constructs composed of HUVEC/HDF cell aggregates had a small deformation percentage 85% after 3-days incubation) and were sufficiently sturdy to be handled and transferred as shown in Fig. 10b.[82]. MOVAS/HUVEC/NIH 3T3 multicellular aggregates fused within 3 days, which corresponds to earlier studies [25, 80]. The cell viability upon implementation was high (97%) showing that the cellular bioink preparation method is successful in comparison to other studies in literature. It seems that multicellular aggregates composed of human cells have better mechanical properties. This research is supported by The Scientific and Technological Research Council of Turkey (TUBITAK) grant number 112M094.

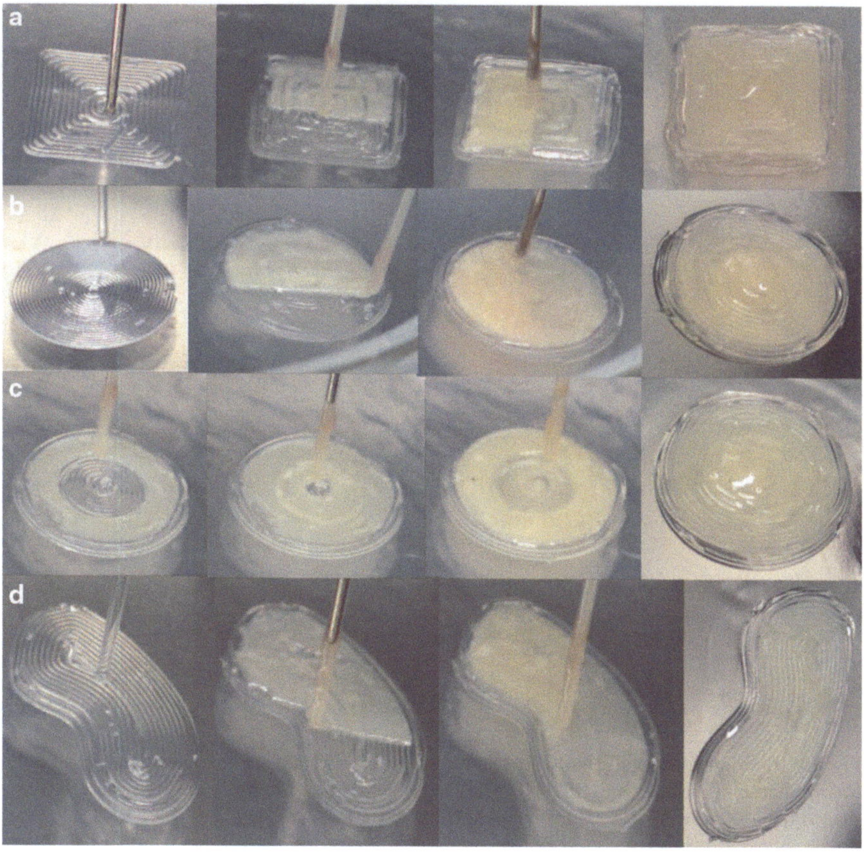

Fig. 9 Continuous bioprinting of live cells directly from computer models **a** *Rectangular* shaped **b** Random-shaped **c** Zig-zag patterned *circular* **d** Spiral patterned circular printed structures [82]

3 Conclusion and Discussion on Stem Cell Printing

Bioprinting is one of the most promising techniques in tissue engineering, where living cells are deposited layer-by-layer with or without biomaterials in user-defined patterns to build 3D tissue constructs. However, there are some challenges related with technical, material and cellular aspects. 3D bioprinting technology requires increased resolution, speed and compatibility with biologically relevant materials. Especially, the fabrication speed must be increased to create structures of clinically relevant sizes. Even the cells used for bioprinting applications are robust enough to survive the bioprinting process and withstand the physiological stresses; a large cell construct in an open environment may not survive a slow and therefore long printing process. For bioprinting, well-characterized and reproducible source of cells is required and any cell type selected for printing should be able to be proliferated into sufficient numbers for printing. Additionally, the proliferation rate and the differentiation with small molecules or other factors should be con-

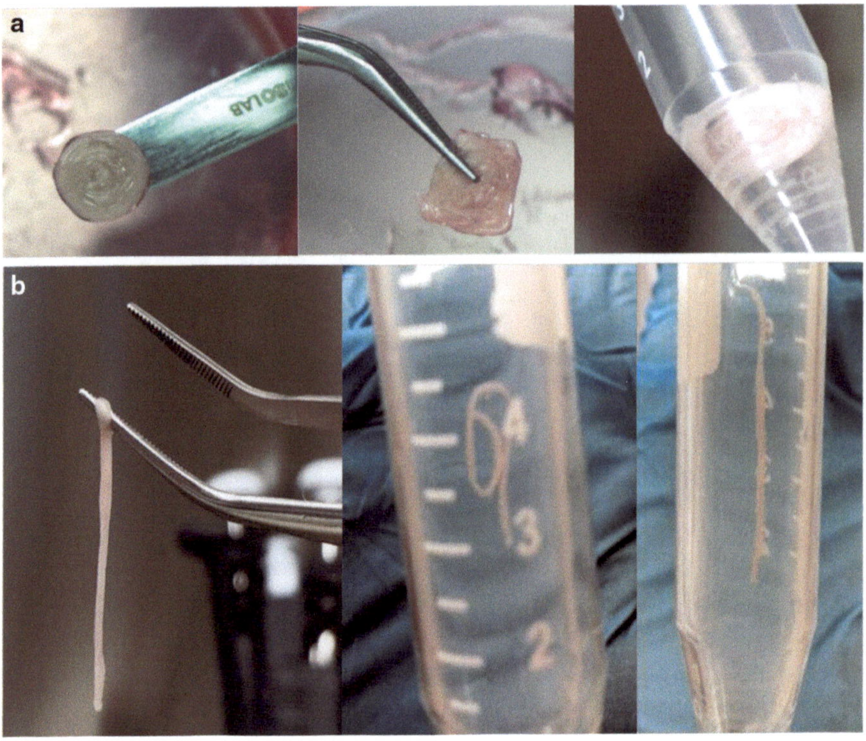

Fig. 10 Bioprinted 3D totally biological tissue constructs without any biomaterial. **a** *Circular* and *square* shaped bioprinted with MOVAS, HUVEC and NIH 3T3 multicellular aggregates. **b** Stripe shaped bioprinted with HUVEC/HDF cell aggregates [82]

trollable. Furthermore, sufficient vascularization and capillaries/microvessels are required for long-term viability of the fabricated construct and for tissue perfusion, respectively. The engineered structure should have suitable mechanical properties for physiological pressures and for surgical connection in case of transplantation. Bioreactors are used to maintain tissues in vitro and to provide maturation factors as well as physiological stressors for assembly, differentiation and ECM production prior to in vivo implantation.

Stem cells such as mesenchymal, induced pluripotent and embryonic stem cells could be a great source for bioprinting. Especially, printing differentiated or progenitor cells precisely and spatially could lead to multi-functional tissue constructs or even organs. However several challenges need to be overcome. First, bioprinting with stem cells requires large number of cells. Culturing this amount of stem cells could be really difficult. Especially, growing large number of stem cells on a feeder layer or special growth medium is really challenging. Even after culturing enough number of stem cells, bioprinting stem cells precisely and spatially accurate manner would require highly precise and special bioprinters. During bioprinting, the effect of compression on the viability and differentiation of stem cells should be considered. Since the stem cells are more susceptible to bioprinting conditions, more gentle and short printing procedures should be developed. Although the current 3D bioprinting technology shows a great deal of promise to generate 3D layered constructs using live mixed cell populations, there is still a long way to go to create a fully-functional organ.

References

1. Langer R, Vacanti JP. Tissue engineering. Science. 1993;260(0036-8075 (Print)):920–6.
2. Pollok JM, Vacanti JP. Tissue engineering. Semin Pediatr Surg. 1996;8586(1055-8586 (Print)):191–6.
3. Nakayama K. In vitro biofabrication of tissues and organs. In: Forgacs G, Sun W, editors. Biofabrication: micro- and nano-fabrication, printing, patterning and assemblies. Amsterdam: Elsevier; 2013. p. 1–16. doi:10.1016/B978-1-4557-2852-7.00013-5.
4. Atala A, Bauer SB, Soker S, Yoo JJ, Retik AB. Tissue-engineered autologous bladders for patients needing cystoplasty. Lancet. 2006;367(9518):1241–6.
5. Brittberg M, Nilsson A, Lindahl A, Ohlsson C, Peterson L.. Rabbit articular cartilage defects treated with autologous cultured chondrocytes. Clin Orthop Relat Res. 1996;326(0009-921X (Print)):270–83.
6. Gallico GG, O'Connor NE, Compton CC, Kehinde O, Green H. Permanent coverage of large burn wounds with autologous cultured human epithelium. New Engl J Med. 1984;311(7):448–51.
7. Hibino N, McGillicuddy E, Matsumura G, Ichihara Y, Naito Y, Breuer C, et al. Late-term results of tissue-engineered vascular grafts in humans. J Thorac Cardiovasc Surg. 2010;139(2):431–6.e2.
8. Matsumura G, Ishihara Y, Miyagawa-Tomita S, Ikada Y, Matsuda S, Kurosawa H, Shin'oka T, et al. Evaluation of tissue-engineered vascular autografts. Tissue Eng. 2006;12(1076–3279 (Print)):3075–83.
9. Niemeyer P, Krause U, Fellenberg J, Kasten P, Seckinger A, Ho AD, Simank HG, et al. Evaluation of mineralized collagen and alpha-tricalcium phosphate as scaffolds for tissue engineering of bone using human mesenchymal stem cells. Cells Tissues Organs. 2004;177(2)(1422–6405 (Print)):880–5.
10. Oberpenning F, Meng J, Yoo JJ, Atala A. De novo reconstitution of a functional mammalian urinary bladder by tissue engineering. Nat Biotech. 1999;17(2):149–55.
11. Radisic M, Park H, Shing H, Consi T, Schoen FJ, Langer R, et al. Functional assembly of engineered myocardium by electrical stimulation of cardiac myocytes cultured on scaffolds. Proc Nati Acad Sci. 2004;101(52):18129–34.
12. Shin'oka T, Matsumura G, Hibino N, Naito Y, Watanabe M, Konuma T, et al. Midterm clinical result of tissue-engineered vascular autografts seeded with autologous bone marrow cells. J Thorac Cardiovascr Surg. 2005;129(6):1330–8.
13. Nakamura M, Iwanaga S, Henmi C, Arai K, Nishiyama Y. Biomatrices and biomaterials for future developments of bioprinting and biofabrication. Biofabrication. 2010;2(1):014110.
14. Wüst S, Müller R, Hofmann S. Controlled positioning of cells in biomaterials—approaches towards 3D tissue printing. J Funct Biomater. 2011;2(3):119–54.
15. Melchels FPW, Domingos MAN, Klein TJ, Malda J, Bartolo PJ, Hutmacher DW. Additive manufacturing of tissues and organs. Prog Polym Sci. 2012;37(8):1079–104.
16. Khoda AKM, Ozbolat IT, Koc B. A functionally gradient variational porosity architecture for hollowed scaffolds fabrication. Biofabrication. 2011;3(3):034106.
17. Khoda AKM, Ozbolat IT, Koc B. Designing heterogeneous porous tissue scaffolds for additive manufacturing processes. Comput-Aided Des. 2013;45(12):1507–23.
18. Matsuda N, Shimizu T, Yamato M, Okano T. Tissue engineering based on cell sheet technology. Adv Mater. 2007;19(20):3089–99.

19. Mironov V, Prestwich G, Forgacs G. Bioprinting living structures. J Mater Chem. 2007;17(20):2054–60.
20. Mironov V, Visconti RP, Kasyanov V, Forgacs G, Drake CJ, Markwald RR. Organ printing: tissue spheroids as building blocks. Biomaterials. 2009;30(12):2164–74.
21. Jakab K, Norotte C, Marga F, Murphy K, Vunjak-Novakovic G, Gabor F. Tissue engineering by self-assembly and bio-printing of living cells. Biofabrication. 2010;2(2):022001.
22. L'Heureux N, Dusserre N, Konig G, Victor B, Keire P, Wight TN, et al. Human tissue-engineered blood vessels for adult arterial revascularization. Nat Med. 2006;12(3):361–5.
23. L'Heureux N, Pâquet S, Labbé R, Germain L, Auger FA. A completely biological tissue-engineered human blood vessel. FASEB J. 1998;12(1):47–56.
24. Ferris C, Gilmore K, Wallace G, in het Panhuis M. Biofabrication: an overview of the approaches used for printing of living cells. Appl Microbiol Biotechnol. 2013;97(10):4243–58.
25. Marga F, Jakab K, Khatiwala C, Shepherd B, Dorfman S, Hubbard B, et al. Toward engineering functional organ modules by additive manufacturing. Biofabrication. 2012;4(2):022001.
26. Forgacs G, Foty RA, Shafrir Y, Steinberg MS. Viscoelastic properties of living embryonic tissues: a quantitative study. Biophys J. 1998;74(5):2227–34.
27. Koç B, Hafezi F, Ozler SB, Kucukgul C. Bioprinting-application of additive manufacturing in medicine. In: Bandyopadhyay A, Bose S, editors. Additive manufacturing. CRC Press; in press.
28. Chang CC, Boland ED, Williams SK, Hoying JB. Direct-write bioprinting three-dimensional biohybrid systems for future regenerative therapies. J Biomed Mater Res B Appl Biomater. 2011;98B(1):160–70.
29. Odde DJ, Renn MJ. Laser-guided direct writing for applications in biotechnology. Trends Biotechnol. 1999;17(10):385–9.
30. Nahmias Y, Schwartz RE, Verfaillie CM, Odde DJ. Laser-guided direct writing for three-dimensional tissue engineering. Biotechnol Bioeng. 2005;92(2):129–36.
31. Zhen M, Russell KP, Qin W, Julie XY, Xiaocong Y, Peng X, et al. Laser-guidance-based cell deposition microscope for heterotypic single-cell micropatterning. Biofabrication. 2011;3(3):034107.
32. Murphy SV, Atala A. 3D bioprinting of tissues and organs. Nat Biotech. 2014;32(8):773–85.
33. Guillotin B, Souquet A, Catros S, Duocastella M, Pippenger B, Bellance S, et al. Laser assisted bioprinting of engineered tissue with high cell density and microscale organization. Biomaterials. 2010;31(28):7250–6.
34. Barron JA, Ringeisen BR, Kim H, Spargo BJ, Chrisey DB. Application of laser printing to mammalian cells. Thin Solid Films. 2004;453–454(0):383–7.
35. Ringeisen BR, Kim H, Barron JA, Krizman DB, Chrisey DB, Jackman S, Auyeung RY, et al. Laser printing of pluripotent embryonal carcinoma cells. Tissue Eng. 2004;(1076–3279 (Print))10(3–4):483–91.
36. Guillemot F, Souquet A, Catros S, Guillotin B, Lopez J, Faucon M, et al. High-throughput laser printing of cells and biomaterials for tissue engineering. Acta Biomater. 2010;6(7):2494–500.
37. Haraguchi Y, Shimizu T, Yamato M, Okano T. Regenerative therapies using cell sheet-based tissue engineering for cardiac disease. Cardiol Res Pract. 2011;2011:845170.
38. Shimizu T, Yamato M, Isoi Y, Akutsu T, Setomaru T, Abe K, Kikuchi A, et al. Fabrication of pulsatile cardiac tissue grafts using a novel 3-dimensional cell sheet manipulation technique and temperature-responsive cell culture surfaces. Circul Res. 2002;90(1524–4571 (Electronic)):e40.
39. Imen Elloumi H, Masayuki Y, Teruo O. Cell sheet technology and cell patterning for biofabrication. Biofabrication. 2009;1(2):022002.
40. Shimizu T, Yamato M, Akutsu T, Shibata T, Isoi Y, Kikuchi A, et al. Electrically communicating three-dimensional cardiac tissue mimic fabricated by layered cultured cardiomyocyte sheets. J Biomed Mater Res. 2002;60(1):110–7.

41. Haraguchi Y, Shimizu T, Yamato M, Kikuchi A, Okano T. Electrical coupling of cardiomyocyte sheets occurs rapidly via functional gap junction formation. Biomaterials. 2006;27(27):4765–74.
42. Shimizu T, Sekine H, Isoi Y, Yamato M, Kikuchi A, Okano T. Long-term survival and growth of pulsatile myocardial tissue grafts engineered by the layering of cardiomyocyte sheets. Tissue Eng. 2006;12(1076–3279 (Print)):499–507.
43. Yamato M, Utsumi M, Kushida A, Konno C, Kikuchi A, Okano T.. Thermo-responsive culture dishes allow the intact harvest of multilayered keratinocyte sheets without dispase by reducing temperature. Tissue Eng. 2001;7(1076–3279 (Print)):473–80.
44. Kushida A, Yamato M, Isoi Y, Kikuchi A, Okano T. A noninvasive transfer system for polarized renal tubule epithelial cell sheets using temperature-responsive culture dishes. Eur Cell Materl 2005;10(1473–2262 (Electronic)):23–30.
45. Akizuki T, Oda S, Komaki M, Tsuchioka H, Kawakatsu N, Kikuchi A, et al. Application of periodontal ligament cell sheet for periodontal regeneration: a pilot study in beagle dogs. J Periodontal Res. 2005;40(3):245–51.
46. Hasegawa M, Yamato M, Kikuchi A, Okano T, Ishikawa I.. Human periodontal ligament cell sheets can regenerate periodontal ligament tissue in an athymic rat model. Tissue Eng. 2005;11(1076–3279 (Print)):469–78.
47. Malda J, Visser J, Melchels FP, Jüngst T, Hennink WE, Dhert WJA, et al. 25th anniversary article: engineering hydrogels for biofabrication. Adv Mater. 2013;25(36):5011–28.
48. Nakamura M, Kobayashi A, Takagi F, Watanabe A, Hiruma Y, Ohuchi K, Iwasaki Y, et al. Biocompatible inkjet printing technique for designed seeding of individual living cells. Tissue Eng. 2005;11(1076–3279 (Print)):1658–66.
49. Wilson WC, Boland T. Cell and organ printing 1: protein and cell printers. Anat Rec A: Discov Mol, Cell, Evol Biol. 2003;272 A(2):491–6.
50. Boland T, Mironov V, Gutowska A, Roth EA, Markwald RR. Cell and organ printing 2: fusion of cell aggregates in three-dimensional gels. Anatomical Rec A: Discov Mol, Cell, Evol Biol. 2003;272 A(2):497–502.
51. Xu T, Jin J, Gregory C, Hickman JJ, Boland T. Inkjet printing of viable mammalian cells. Biomaterials. 2005;26(1):93–9.
52. Cui X, Boland T. Human microvasculature fabrication using thermal inkjet printing technology. Biomaterials. 2009;30(31):6221–7.
53. Boland T, Tao X, Damon BJ, Manley B, Kesari P, Jalota S, et al. Drop-on-demand printing of cells and materials for designer tissue constructs. Mater Sci Eng: C. 2007;27(3):372–6.
54. Nakamura M, Nishiyama Y, Henmi C, Yamaguchi K, Mochizuki S, Koki T, et al. Inkjet bioprinting as an effective tool for tissue fabrication. NIP Digital Fabr Conf. 2006;2006(3):89–92.
55. Nishiyama Y, Nakamura M, Henmi C, Yamaguchi K, Mochizuki S, Nakagawa H, et al. Development of a three-dimensional bioprinter: construction of cell supporting structures using hydrogel and state-of-the-art inkjet technology. J Biomech Eng. 2008;131(3):035001.
56. Xu C, Chai W, Huang Y, Markwald RR. Scaffold-free inkjet printing of three-dimensional zigzag cellular tubes. Biotechnol Bioeng. 2012;109(12):3152–60.
57. Khatiwala C, Law R, Shepherd B, Dorfman S, Csete M. 3D cell bioprinting for regenerative medicine research and therapies. Gene Ther Regul. 2012;07(01):1230004.
58. Shabnam P, Madhuja G, Frédéric L, Karen CC. Effects of surfactant and gentle agitation on inkjet dispensing of living cells. Biofabrication. 2010;2(2):025003.
59. Chahal D, Ahmadi A, Cheung KC. Improving piezoelectric cell printing accuracy and reliability through neutral buoyancy of suspensions. Biotechnol Bioeng. 2012;109(11):2932–40.
60. Ferris CJ, Gilmore KJ, Beirne S, McCallum D, Wallace GG, In het Panhuis M. Bio-ink for on-demand printing of living cells. Biomater Sci. 2013;1(2):224–30.
61. Foty RA, Steinberg MS. The differential adhesion hypothesis: a direct evaluation. Dev Biol. 2005;278(1):255–63.
62. Steinberg MS. Differential adhesion in morphogenesis: a modern view. Curr Opin Genet Dev. 2007;17(4):281–6.

63. Steinberg MS. Reconstruction of tissues by dissociated cells. Some morphogenetic tissue movements and the sorting out of embryonic cells may have a common explanation. Science. 1963;141(0036-8075 (Print)):401–8.
64. Jakab K, Norotte C, Damon B, Marga F, Neagu A, Besch-Williford CL, Kachurin A, et al. Tissue engineering by self-assembly of cells printed into topologically defined structures. Tissue Eng. 2008;14(1937–3341 (Print)):413–21.
65. Kelm JM, Lorber V, Snedeker JG, Schmidt D, Broggini-Tenzer A, Weisstanner M, et al. A novel concept for scaffold-free vessel tissue engineering: Self-assembly of microtissue building blocks. J Biotechnol. 2010;148(1):46–55.
66. Gwyther TA, Hu JZ, Christakis AG, Skorinko JK, Shaw SM, Billiar KL, et al. Engineered vascular tissue fabricated from aggregated smooth muscle cells. Cells Tissues Organs. 2011;194(1):13–24.
67. Adebayo O, Hookway TA, Hu JZ, Billiar KL, Rolle MW. Self-assembled smooth muscle cell tissue rings exhibit greater tensile strength than cell-seeded fibrin or collagen gel rings. J Biomed Mater Res A. 2013;101 A(2):428–37.
68. Napolitano AP, Chai P, Dean DM, Morgan JR. Dynamics of the self-assembly of complex cellular aggregates on micromolded nonadhesive hydrogels. Tissue Eng. 2007;13(1076–3279 (Print)):2087–94.
69. Dean DM, Napolitano AP, Youssef J, Morgan JR. Rods, tori, and honeycombs: the directed self-assembly of microtissues with prescribed microscale geometries. FASEB J. 2007;21(14):4005–12.
70. Khalil S, Nam J, Sun W. Multi-nozzle deposition for construction of 3D biopolymer tissue scaffolds. Rapid Prototyp J. 2005;11(1):9–17.
71. Khalil S, Sun W. Biopolymer deposition for freeform fabrication of hydrogel tissue constructs. Mater Sci Eng: C. 2007;27(3):469–78.
72. Smith CM, Christian JJ, Warren WL, Williams SK. Characterizing environmental factors that impact the viability of tissue-engineered constructs fabricated by a direct-write bioassembly tool. Tissue Eng. 2007;13(1076–3279 (Print)):373–83.
73. Smith CM, Stone AL, Parkhill RL, Stewart Rl, Simpkins MW, Kachurin AM, Warren WL, et al. Three-dimensional bioassembly tool for generating viable tissue-engineered constructs. Tissue Eng. 2004;10(1076–3279 (Print)):1566–76.
74. Duan B, Hockaday LA, Kang KH, Butcher JT. 3D Bioprinting of heterogeneous aortic valve conduits with alginate/gelatin hydrogels. J Biomed Mater Res A. 2012;101 A(5):1255–64.
75. Hockaday LA, Kang KH, Colangelo NW, Cheung PYC, Duan B, Malone E, et al. Rapid 3D printing of anatomically accurate and mechanically heterogeneous aortic valve hydrogel scaffolds. Biofabrication. 2012;4(3):035005.
76. Chang R, Nam J, Sun W. Effects of dispensing pressure and nozzle diameter on cell survival from solid freeform fabrication-based direct cell writing. Tissue Eng. 2008;14(1937–3341 (Print)):41–8.
77. Cohen DL, Malone E, Lipson H, Bonassar LJ. Direct freeform fabrication of seeded hydrogels in arbitrary geometries. Tissue Eng. 2006;12(1076–3279 (Print)):1325–35.
78. Christopher MO, Francoise M, Gabor F, Cheryl MH. Biofabrication and testing of a fully cellular nerve graft. Biofabrication. 2013;5(4):045007.
79. Mironov V, Kasyanov V, Markwald RR. Organ printing: from bioprinter to organ biofabrication line. Curr Opin Biotechnol. 2011;22(5):667–73.
80. Norotte C, Marga FS, Niklason LE, Forgacs G. Scaffold-free vascular tissue engineering using bioprinting. Biomaterials. 2009;30(30):5910–7.
81. Kucukgul C, Ozler SB, Inci I, Karakas E, Irmak S, Gozuacik D, et al. 3D bioprinting of biomimetic aortic vascular constructs with self-supporting cells. Biotechnol Bioeng. 2014:112(4):811–21 n/a-n/a.
82. Ozler SB, Kucukgul C, Koc B. 3D continuous cell bioprinting. submitted 2015.
83. Kucukgul C, Ozler B, Karakas HE, Gozuacik D, Koc B. 3D hybrid bioprinting of macrovascular structures. Procedia Eng. 2013;59(0):183–92.

Hydrogels for Cell Encapsulation and Bioprinting

Seyed Ramin Pajoum Shariati, Seyedsina Moeinzadeh and Esmaiel Jabbari

1 Introduction

Bioprinting is an emerging technology based on a computer-aided design (CAD) approach, in which the controlled deposition of bio-ink, cells with or without hydrogels, on hydrogel materials (as a bio-paper) is achieved by printers inspired by common inkjet printers [1]. Bioprinting combines biology and engineering in order to construct complex structures for tissue engineering applications [2, 3], as well as drug screening and toxicology [4, 5]. Bioprinters can fabricate 3D biomimetic tissues in a spatially controlled manner with high precision over the shape, size, cell location and enable the fabrication of functional organs comprising of multiple cell types by providing similar cellular microenvironments to those found *in vivo* [1, 6, 7]. Bioprinters can be classified into two groups. The first group is nozzle based which can be further divided into intermittent drop-wise printers- such as inkjet printers (both thermal and piezoelectric)—and continuous robotic dispensing printers. The second group is drop-wise nozzle-free which is based on laser-induced forward transfer printing techniques (Fig. 1) [8].

Hydrogels have been used in numerous tissue engineering approaches including rapid prototyping techniques. Bioprinting is another application of hydrogels based on their potential for entrapping living cells and their ease of processing [9]. Hydrogels play a significant role in bioprinting as an indispensable matrix material

E. Jabbari (✉)
Chemical and Biomedical Engineering, Swearingen Engineering Center, University of South Carolina, Rm 2C11, Columbia, SC 29208, USA
e-mail: jabbari@mailbox.sc.edu

S. R. Pajoum Shariati · S. Moeinzadeh · E. Jabbari
Biomimetic Materials and Tissue Engineering Laboratory, Department of Chemical Engineering, University of South Carolina, Columbia, SC 29208, USA

K. Turksen (ed.), *Bioprinting in Regenerative Medicine,* Stem Cell Biology and Regenerative Medicine, DOI 10.1007/978-3-319-21386-6_4

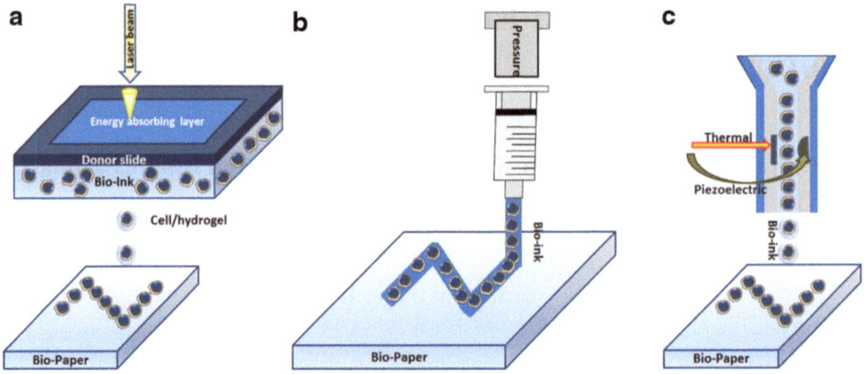

Fig. 1 Different approaches used for bioprinting with a bio-ink composed of cells encapsulated in a hydrogel: **a** laser-induced forward transfer, **b** robotic dispensing, and **c** inkjet printing

for direct fabrication of living cells into tissue-like structures. The required hydrogel characteristics depend on the type of bioprinter and the printing approach. For example, bioprinters based on laser-induced forward transfer technique do not have a nozzle, allowing for a wide range of hydrogel viscosities since clogging is not an issue in this approach. On the other hand, the above-mentioned approach needs rapid gelation kinetics and relatively low flow rate for the bio-ink to achieve high shape reproducibility; this limits the construction of large 3D structures for clinical applications [10]. Clogging is an important issue in bioprinters based on inkjet technique. Inkjet bioprinters require low hydrogel viscosity for the bio-ink which limits the use of many viscous natural materials and fabrication of 3D structures [11]. In contrast to inkjet printers, robotic dispensing techniques use hydrogels with higher viscosities, making them suitable for fabrication of structures with clinically relevant size, but the technique is limited by lower resolution. [12]

Various types of hydrogels with different mechanisms for crosslinking have been used for bioprinting. Some of these hydrogels include collagen type I [13–15], collagen/fibrin [15, 16], fibrin [15, 17], Extracel™ Hydrogel [15], hyaluronic acid (HA) [1, 15, 18], hyaluronan-based hydrogels [19, 20], hyaluronic acid and dextran-based semi-interpenetrating networks (semi-IPN) [21], tyramine-substituted hyaluronic acid (TS-NaHy) [15], Corgel™ [15], methylcellulose-hyaluronan (MC-HA) [15], chitosan [15], chitosan/collagen [15], methacrylated gelatin (GelMA) [22], polyethylene glycol diacrylate (PEGDA) [1, 15], agarose [15, 23], alginate [7, 15, 24–26], alginate/gelatin [15, 27], PEG [1], polyacrylamide-based hydrogels [28], and NovoGel [29]. Many factors need to be considered when selecting a hydrogel for bioprinting such as cytotoxicity, gelation time, extent of swelling, viscosity, printability, and hydrogel stability [15]. Cell death in the center of the printed hydrogel [30, 31], due to the insufficient depth of penetration of oxygen and nutrients, should be considered [32]. Selection of hydrogels for bioprinting also depends on the target tissue. Bioprinting can be used to fabricate many different tissues such as skin [15], nerve [16, 28, 33], musculoskeletal [21, 25], vasculature [19, 20, 29, 33], and heart [27].

Hydrogels can be used as bio-inks as well as bio-papers in bioprinting. In this chapter, we will discuss the required properties of hydrogels and their characteristics including viscosity, gelation time, swelling, degradation and mechanical properties for bioprinting applications.

2 Hydrogels as Bio-Ink

The hydrogel precursor solution containing cells and cell aggregates is considered the bio-ink for bioprinting. Since mammalian cells are very sensitive to heat and mechanical stress, special steps should be taken in order to avoid cell damage and lysis. Moreover, their physiological state and function depend on their environment and culture conditions. Hydrogels can mitigate the harsh conditions of bioprinting such as the high temperature employed at the nozzle by thermal printers [11]. Furthermore, cell-laden hydrogels as building blocks of tissue-engineered constructs can protect the cells from the host immune system [26]. For example, xenogenic endocrine cells are encapsulated and transplanted into patients who suffer from diabetes [34], anemia [35] and dwarfism [36].

In the design of a two-component bio-ink (cells and hydrogel), fast-gelling polymerization reactions should be considered. This can be achieved by using the alginate/Ca^{2+} reaction [37–39], the fibrin/thrombin reaction [17, 40, 41], or photopolymerizable inks [42]. Alginate is widely used for cell encapsulation because of its fast sol-gel transition either by mixing cells with alginate and printing the cell precursor solution into a crosslinking Ca^{2+} solution [37, 38] or by mixing cells with Ca^{2+} and printing the cell suspension into an alginate or alginate/collagen solution [39]. The surface of an alginate droplet is crosslinked by divalent cations (Ca^{2+}) resulting in an increased viscosity and bonding of the droplet to the surface [26]. Fibrin is a versatile biopolymer formed during blood clot formation by the enzymatic reaction of fibrinogen with thrombin (the enzyme which catalyzes the polymerization of fibrinogen) in the presence of calcium ions. The suspension of cells in a fibrin hydrogel has been used as a bio-ink for laser assisted [17] and thermal inkjet [41] bioprinting. Since fibrin has a long gelation time (5–15 min), a mixture of fibrin and alginate biopolymers has been proposed to accelerate gelation using a 3D plotter in two steps involving alginate/Ca^{2+} gelation followed by fibrinogen/thrombin reaction [43].

Photocrosslinkable polyethylene glycol (PEG) based hydrogels have also been widely used for cell encapsulation. Acrylated-functionalized PEGs of different molecular weights are often used to produce photocrosslinkable hydrogels. Since acrylated PEGs are not biodegradable, chemically-modified biodegradable PEGs have been synthesized for bioprinting [44]. Further, the cell adhesion characteristics of PEG hydrogels can be improved by functionalization with biomolecules, proteins [45, 46] as well as peptides, such as arginine-glycine-aspartic acid (RGD) [47]. Other photocrosslinkable polymers include poly(ethylene oxide) dimethacrylate [48], photocrosslinkable alginate hydrogels with tunable degradation and mechanical properties [49], and photocrosslinkable Lutrol hydrogel [50]. A jet of beads can be formed with some of these polymers [51] and the sol-gel transition in some cases is relatively fast in the order of 15 s [52] which makes these hydrogels suitable for bioprinting.

The above-mentioned hydrogels cannot be applied to every bioprinting situation as the application depends on cell type and target tissue. Therefore, smarter bio-inks are needed in the future to facilitate the drop-on-demand cell printing. Printability, maintaining the cells in their functional and physiological conditions, and optimal fluid properties to prevent cell settling and aggregation are the most challenging tasks for developing a bio-ink formulation. In that regard, a surfactant-containing gellan gum has been used as a bio-ink to address the issue of facile cell deposition for drop-on-demand printing using both a commercial micro-valve deposition system, and a multi-nozzle piezoelectric inkjet print head [53]. This drop-on-demand based approach achieved high cell viability especially for sensitive cell types. However, the cell encapsulation is random in this approach as a cell suspension reservoir is used. Thus, this approach can only be used after cell sorting [54].

Aside from individual cells suspended in hydrogels, cell aggregates suspended in gels have also been used as bio-ink [55]. The concept of self-assembling bio-ink and cell aggregates as a building block is especially useful in organ printing where both cell number density and the complexity of fabrication are increasingly important [55]. Advantages of cell aggregates over individual cells in bioprinting include a) facile fabrication of 3D structures by fusion of small pre-formed tissue blocks of cell aggregates, b) decrease in printing time due to the increased cell number density, and c) increase in cell survival due to the mild shear stresses exerted by micropipettes in printing cell aggregates compared to the stresses for single cells. The hydrogels used with cell aggregate should provide favorable conditions for self-assembly after bioprinting in addition to the material requirements mentioned earlier for single cells. The hydrogel for printing cell aggregates can be selected from thermo-reversible or thermo-sensitive gels [56–62], photo-sensitive gels [18, 63–66], pH-sensitive gels [16, 67], and gels sensitive to specific factors in the microenvironment [68]. Future studies should lead to the developments of new hydrogel precursors as bio-inks with fast and efficient gelation kinetics that can satisfy the requirements of the target tissue.

3 Hydrogels as Bio-Paper

Bio-paper can be defined as a biocompatible, degradable, in situ crosslinkable, and biomimetic hydrogel that is permissive to tissue fusion in order to serve as the substrate for cell deposition in bioprinting (Fig. 2) [69]. Based on the above definition, there are distinct differences between the hydrogels used in tissue engineering and those used in bioprinting. The bio-paper provides a temporary support for the cells printed with the bio-ink but it should be possible to remove the bio-paper

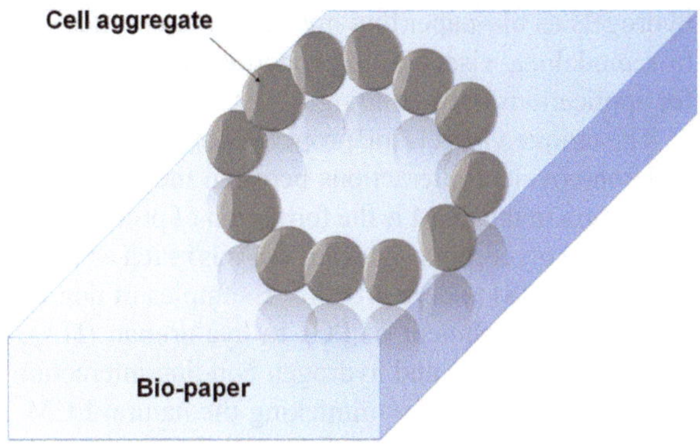

Fig. 2 Schematic diagram of a bio-paper serving as a substrate for assembly and fusion of cell aggregates into a defined structure

after tissue printing. Hydrogels that are stimuli-sensitive and in situ crosslinkable and rapidly solidify after injection are often used as the bio-paper. With regard to crosslinking, the viscosity of the hydrogel precursor solution should be sufficiently low to allow injection but it should solidify immediately after injection to facilitate cell deposition. In addition to those requirements, the gel should be sufficiently stiff to maintain stability and shape after deposition of cells or cell aggregates on the bio-paper.

Another important property of the bio-paper is the ability to resist contraction or molecular stenting. The bio-paper should maintain its shape and resist the cell-generated contractile force when cells are deposited on its surface [23]. In the native extracellular matrix, structural stability and viscoelastic properties of the network of collagen fibers, the most abundant protein of ECM, are reinforced by aggrecan and versican [70]. In addition to mechanical stability, the hydrogel as a bio-paper should be biocompatible and guide cell adhesion, proliferation, and differentiation to the proper lineage, leading to the formation of a 3D tissue. The hydrogel should also prevent the deposited cells from drying and prevent bleeding of the bio-ink from cell aggregates. The role of hydrogel as the bio-paper is similar to that of the ECM, which is to provide a microenvironment for cell growth and control cell fate by directing cell proliferation, differentiation, and apoptosis. ECM proteins and growth factors can be added to the hydrogel precursor solution to improve cell adhesion or control cell function. In addition, ECM proteins or growth factors can be patterned on the hydrogel to control spatial organization and differentiation of the cells printed on the bio-paper [71–76]. As an example, the hydrogel can be conjugated with integrin-binding arginine-glycine-aspartic acid (RGD) ligands at optimal concentration to improve cell adhesion but not too high to adversely affect cell migration. Intermediate adhesion is sufficient for cell adhesion and migration [77]. Fibronectin [78] and heparin-binding peptides [79] are examples of other ECM molecules that have been used to improve cell adhesion to hydrogels. In addition to the above requirements, mass transport properties of the hydrogel bio-paper which are affected by crosslink density and gel microstructure should be carefully considered to optimize viability of the seeded cells [80]. There is a need to develop new and improve hydrogels as bio-paper that mimics the properties of the natural ECM with respect to remodeling, viscoelastic properties, mechanical stability, and biocompatibility for applications in bioprinting.

The unique viscoelastic properties of the natural ECM are rooted in the covalent and non-covalent interactions between the biopolymers. One example of covalent interaction in the ECM is the formation of proteoglycans (PGs) by covalent bonding of sulfated glycosaminoglycans (GAGs) such as chondroitin sulfate (CS) and heparan sulfate (HS) to core proteins. Examples of non-covalent secondary interactions include the attachment of PGs to hyaluronan (HA), hydration of polysaccharide chains, electrostatic and hydrogen bonding interaction between the collagen chains and collagen fibrils. By mimicking the natural ECM, synthetic biocompatible and in situ crosslinkable hydrogels based on GAGs have been developed [81] that could potentially be applied to bioprinting. These synthetic ECM-like hydrogels have shown satisfactory performance both *in vitro* and *in vivo* especially with regard to spatiotemporal control of cell function and molecular stenting [70]. Prior to using the ECM-like gels as bio-paper, their rheological properties should be evaluated with respect to injectability.

Another property of the hydrogel that should be considered is providing a permissive environment for the fusion of neighboring cell aggregates in order to form a construct with the desired geometry [55, 82]. The concept of cell fusion originally developed for embryonic tissues [83] is based on the fact that cell aggregates can be considered as fluids or liquid [55]. Since the printed cell aggregates can collapse into a large single spherical aggregate to attain the lowest energy state, the rate of cell movement should be controlled to avoid such a collapse [84]. Therefore, the desired 3D cellular structure should be maintained long enough to allow the removal of bio-paper immediately after fusion. Cell-gel interfacial parameters, such as interfacial tension which is related to the density and nature of cell–cell [85, 86] and cell–matrix adhesion molecules [87] determine the permissive ability of the hydrogel for cell fusion. Desirable properties can be achieved by altering the interfacial tension of bio-paper and bio-ink. The relationship between the magnitude of cell-cell and cell-gel interactions determine the life-time of structures printed onto the bio-paper [55, 88]. The structure formed after bioprinting is maintained by molecular mechanisms underlying cell–cell and cell–matrix adhesion as well as cell motility.

There are very few *in vitro* methods to quantify cell fusion [89]. One method measures the angle between the tangents to the projected boundary of adjacent cell aggregates. Computational methods have also been used to predict the ability of hydrogels to aid cell fusion [55, 82]. Based on the results of computational models, the formation of larger structures is based on the liquidity (water content and extent of swelling) of bio-paper and cell aggregates printed with the bio-ink [55, 82]. Depending on the tissue geometry desired, bio-papers can be sheet-like or rod-like which can be removed to form a flat or tubular structure, respectively [33]. Understanding and controlling cell fusion can allow the printing of vascularized organs with intricate shapes and different cell types [33]. In this regard, the use of preformed cell aggregates that undergo fusion is preferred over individual cells [90].

Bio-papers can be selected from natural as well as synthetic hydrogels or their mixture. Naturally-derived biopolymers which have been used as a bio-paper include collagen [40, 53, 62, 84, 91–93], soy agar and rat-tail type I collagen [40, 94], agar [94], agarose [33], agarose and collagen [13], fibrinogen [17, 40, 41], Matrigel™ [17, 92], and alginate [95]. Synthetic ECM-like gels used as bio-paper include a mixture of thiolated dithiopropionylhydrazide (DTPH), modified gelatin (Gtn-DTPH) and hyaluronan (HA-DTPH) [96, 97], copolymer of polyethylene glycol diacrylate (PEGDA) and Gtn-DTPH [23], gelatin mixed with RGD peptide [98] or a mixture of three recombinant domains of human fibronectin [99], mixture of PVA, hyaluronan and $CaCl_2$ [37], copolymer of N-isopropylacryamide-co-2-(N,N-dimethylamino)-ethyl acrylate [62], hyaluronan hydrogel crosslinked with polyethylene glycol tetra-acrylate [20], methacrylated HA and gelatin [18], and polyacrylamide-based hydrogel [28].

4 Properties of Bioprinting Hydrogels

4.1 Viscosity

A high viscosity for the bio-ink prevents droplet formation at the nozzle but increases the resolution of the printed structure [100]. In addition, an increase in the viscosity of the bio-ink decreases spreading of the ejected droplet on the substrate prior to stabilization by crosslinking [8]. On the other hand, an increase in the viscosity of bio-ink increases the applied shear stress during injection which can damage the encapsulated cells [8]. Therefore, there is an upper limit of 0.1 $Pa.s^{-1}$ for the viscosity of a bio-ink in inkjet printers [11, 101]. For example, relatively viscous solutions of alginate, gelatin-alginate, chitosan, and chitosan-collagen cannot be injected with inkjet bioprinters [15]. The viscosity of a hydrogel precursor solution typically increases with increasing the polymer concentration, polymer molecular weight and attractive inter-molecular interactions between polymer chains [102]. An increase in the polymer concentration in the bio-ink leads to a significant increase in cross-link density and decreases the mesh size and water content of the printed cell-laden hydrogel, which can negatively affect cell motility and proliferation [103, 104]. As a result, polymers with higher molecular weights or polymers capable of inter-molecular physical interaction are typically used to increase the bio-ink viscosity [8, 105]. Further, the molecular weight of the polymer and the viscosity of the bio-ink can be increased via partial crosslinking of the polymer before bio-printing [18, 106]. For example, the viscosity of a photocrosslinkable hyaluronan-gelatin bio-ink was increased by UV irradiation of the precursor solution for 120 s to produce a gel-like fluid prior to bio-printing [18].

4.2 Gelation Time

Bio-printing hydrogels should have a fast gelation time, typically in the order of 2 min, to ensure stability of the printed structure after deposition and before post-crosslinking [8, 15]. The gelation time of natural hydrogels with slow crosslinking rates can be decreased by mixing with a fast crosslinking macromer or by chemical modification of physical gels with covalent bonds [15]. For example, the gelation time of collagen decreased from 40 min to 15 s by mixing with fibrin gel [15] and the gelation time of Extracel hydrogel (Glycosan Biosystems, Alameda, CA) decreased from 30 min to 18 s by incorporation of UV initiated covalent crosslinking concurrent with physical crosslinking [15].

The gelation kinetics and viscoelastic behavior of physically and chemically crosslinked gels can be evaluated by rheometry. The gelation kinetics of physical micellar gels depends on the strength and number density of physical crosslinks [105, 107]. The gelation kinetics of covalently-bonded chemical gels depends on several factors including the concentration and distribution of crosslinkable groups and initiator [104, 108–110]. Figure 3a shows the gelation time of photocrosslinkable star poly(ethylene glycol-co-lactide) acrylate (SPELA) macromers with respect to concentration of UV photo-initiator. As the initiator concentration increased from 0.08 to 0.78 wt%, gelation time of the macromers decreased from 200±9 to 42±2 s [104]. A decrease in gelation time by increasing the concentration of photo-initiator was related to an increase in the propagation rate of crosslinks by

$$R_P = K_P[AC]\left[\frac{\varphi \varepsilon I_0 \delta[I]}{K_t}\right]^{1/2} \tag{1}$$

where K_P and K_t are the rate constants for chain propagation and termination respectively, $[AC]$ is the concentration of unreacted acrylates, φ is initiation efficiency, ε is molar extinction coefficient, I_0 is the intensity of incident radiation, δ is sample thickness, and $[I]$ is photo-initiator concentration [111]. According the

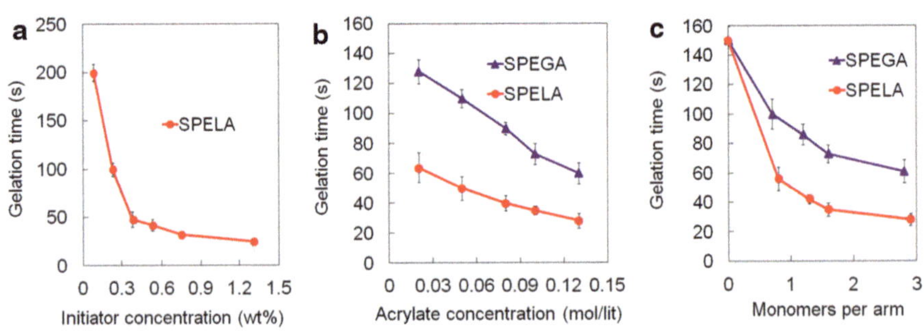

Fig. 3 Effect of initiator concentration **a**, on gelation time of *SPELA* hydrogel (20 wt%, m = 1.7). Effect of acrylate concentration **b**, and number of monomers per macromer arm **c**, on gelation time of *SPELA* and *SPEGA* hydrogels

above equation, the propagation rate of crosslinks for SPELA macromers increased with $[I]$ leading to a decrease in gelation time (Fig. 3a). Figure 3b compares the gelation time of star poly(ethylene glycol-co-glycolide) acrylate (SPEGA) with that of SPELA macromer as a function of the concentration of reactive acrylate groups. The gelation time of SPEGA hydrogel decreased from 128 to 60 s while that of SPELA decreased from 64 to 28 s with increasing acrylate concentrations from 0.02 to 0.13 mol/L.

A decrease in gelation time with increasing $[AC]$ concentration was attributed to an increase in the rate of propagation reaction (see Eq. 1). Dissipative Particle Dynamics (DPD) simulations showed that a large fraction of the crosslinking reaction occurred in the aqueous phase for the less hydrophobic SPEGA macromer whereas a large fraction of the crosslinking occurred in the micellar phase for the more hydrophobic SPELA [105, 109]. Therefore, SPELA micellar gels had a faster crosslinking and a shorter gelation time compared to SPEGA gels (see Fig. 3b) [108].

Figure 3c shows the effect of number of hydrophobic lactide/glycolide monomers per macromer (m) on gelation time of the macromers. The gelation time of SPEGA and SPELA macromers decreased sharply from 150 s to 61 and 28 s, respectively, by increasing m from 0 to 3 [108] which was attributed to a change in the distribution of reactive acrylate groups within the hydrogel precursor solution concurrent with micelle formation. The acrylates were positioned in the core of the micelles concurrent with micelle formation in SPEGA/SPELA aqueous solutions. Localization of acrylates within the micelles' core decreased gelation time with increasing m for SPELA/SPEGA macromers (see Fig. 3c). Simulation results have also shown that the proximity of acrylates within the hydrogel precursor solution increased with changing the macromer type from SPEGA to SPELA at the same m value concurrent with micelle formation. These results are consistent with the experimentally-measured shorter gelation times of SPELA macromers compared to those of SPEGA [108].

4.3 Water Content and Swelling

An increase in the degree of swelling or water content typically increases biocompatibility of the bioprinted hydrogel however it adversely affects mechanical properties by increasing the extent of deformation of the printed structure. Furthermore, dissimilar extent of swelling of bio-inks in applications where more than one bio-ink is used can affect the shape of the final construct [8]. In addition, excessively swollen or contracted bio-printed structures may contribute to dehydration of the surrounding tissue or increased scarring in wound healing applications [15, 112]. The degree of swelling and mesh size of the hydrogel are particularly important variables for controlling the diffusion of nutrients, oxygen, growth factors and waste products into or out of the cell-laden hydrogels in tissue engineered constructs [113, 114]. Therefore, the degree of swelling should be optimized for cell-laden bio-printed hydrogels.

The degree of swelling of a hydrogel is controlled by two opposing forces: the thermodynamic force of mixing between the polymer and water and the elastic force of the extended network chains. The force of mixing tends to increase the water content of the gel by attractive interactions between water molecules and the polymer chains [102]. The elastic retractive force of the network chains on the other hand tends to decrease the water content of the gel due to the extension of random-coil chain conformations. The force of mixing and the retractive force of extended chains become equal at equilibrium swelling. Several factors including molecular weight, hydrophobicity, functionality and flexibility of the macromers, degree of crosslinking, nanostructure of the gel, and temperature affect the degree of swelling of the hydrogel [115, 116]. More specifically, the degree of swelling decreases with increasing the extent of crosslinking of the hydrogel and increasing the hydrophobicity of the chains by affecting the elastic energy of the chains and the energy of mixing, respectively [117]. Based on the theory of rubber elasticity and thermodynamics of mixing, the mesh size (ξ) of a hydrogel increases with decreasing crosslink density and increasing the degree of swelling (see Fig. 4a) [117]. For example, the swelling ratio (weight of water divide by the dry gel weight) of SPEGA and SPELA hydrogels decreased from 830 to 430% and from 730 to 370%, respectively, by increasing the reactive acrylate concentration from 0.02 to 0.13 mol/L (Fig. 4b) [118]. The relatively lower swelling ratio of SPELA compared to that of SPEGA can be attributed to the higher hydrophobicity of lactide segments in SPELA compared with glycolide segments in SPEGA (Fig. 4b) [118].

4.4 Degradation

Although the microenvironment of encapsulated cells can be controlled by many factors in the hydrogel structure, the use of hydrogels as a supporting matrix in regenerative medicine is limited by their persistence in the site of regeneration [103].

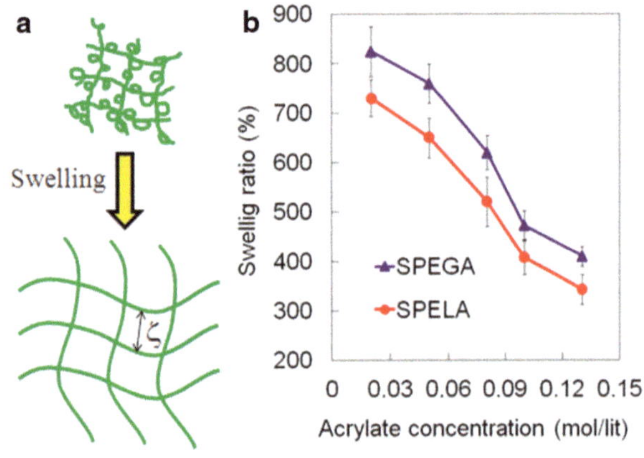

Fig. 4 **a** Schematic representation of the change in mesh size of un-swollen and swollen crosslinked network. **b** Effect of acrylate concentration on the swelling ratio of *SPEGA* and *SPELA* (m = 1.7) hydrogels

Aside from degradation, cell proliferation and migration requires a matrix capable of remodeling. Therefore, the degradation kinetics of bioprinted hydrogels as an engineered matrix should match the rate of tissue formation and the rate of tissue remodeling [103]. Further, the fate of encapsulated cells strongly depends on the extent of degradation and the kinetics of degradation. As an example, mesenchymal stem cells (MSCs) encapsulated in a non-degradable hyaluronic acid gel underwent adipogenic differentiation whereas those encapsulated in a degradable gel underwent differentiation to the osteogenic lineage [119]. As another example, a slow-degrading hydrogel (>50% degradation in 4 weeks) enabled osteogenic differentiation of MSCs whereas a fast-degrading gel (>50% degradation in 1 week) enabled vasculogenic differentiation of endothelial progenitor cells [120–122].

Natural hydrogels that are used as bio-ink degrade enzymatically [8, 15, 101, 123]. Hyaluronic acid based bio-inks degrade by hyaluronidase enzyme [124]. Gelatin based bio-inks completely degrade by 6 h or more depending on enzyme concentration in collagenase solution [121]. Myo-fibroblast cell-laden fibrin gels degrade after 2 days in the absence of inhibitor by enzymatic degradation [125]. Alginate gels can be degraded by alginase enzyme released from PLGA microspheres and the gel degradation rate increased with increasing the alginase loading in microspheres [126]. Synthetic hydrogels on the other hand can be made degradable by conjugation of enzymatic, hydrolytic, or photolytically degradable segments in the macromer or crosslinker chains [127]. In one example, a non-degradable PEG hydrogel was made enzymatically degradable by conjugation of a matrix metalloproteinase (MMP) sensitive peptide to the hydrogel network using the Michael addition reaction [128, 129]. The MMP sensitive PEG gel enhanced migration of fibroblasts encapsulated in the hydrogel compared to the non-degradable PEG gel [128]. Hydrogels based on copolymers of PEG and ε-caprolactone were shown to be hydrolytically degradable but the rate of degradation was limited by the hydrophobicity and phase separation of ε-caprolactone segments in aqueous solution [130]. Copolymerization of PEG with a degradable polymer like poly(lactide) (PLA) has been used to impart degradability to PEG hydrogels. The degradation and water content of the copolymers can be adjusted by the fraction of hydrophobic lactide segments [131, 132] but solubility of the copolymer in aqueous solution was adversely affected by increasing the lactide content [133]. In addition, other hydrolytically degradable polymers including polyphosphoesters [134, 135], poly(ester amides) [136] and poly(amino-ester urethanes) [137] have been used to synthesize degradable hydrogels. Degradable micellar hydrogels with a wide range of degradation rate have been synthesized using star-shape PEG macromers chain-extended with short hydroxy acid segments [104, 108, 118].

4.5 Mechanical Properties

There is a trade-off between the structural stability and biocompatibility of the bioprinted hydrogel [101]. Structural stability, hence mechanical properties, of the bioprinted hydrogel should be optimized to prevent deformation or rupture of the printed construct while maintaining viability and function of the encapsulated cells [138]. Techniques for measuring the mechanical properties of hydrogels for bioprinting include the tensile test or extensometry [139], un-confined compression test [140], confined compression test [141] and indentation test [142]. The latter technique, indentation, is preferred for cell-encapsulated hydrogels because the test can be performed while the construct is immersed in the medium [142–144]. It should be noted that the Young's modulus (E) of the gel is closely related to the indentation modulus [142]. It is well established that matrix stiffness affects the fate of the encapsulated cells. For example, MSCs encapsulated in an alginate hydrogel with 2.5-5 kPa elastic modulus underwent differentiation to adipogenic lineage whereas the same cells encapsulated in a 11–30 kPa alginate gel differentiated to the osteogenic lineage [145]. Physically crosslinked gels due to their low stiffness and elasticity are limited in practical applications by soft tissue compression [104, 146, 147]. Mechanical properties of physical gels can be improved by incorporation of covalent bonds within the hydrogel network [148]. Covalently crosslinked gels have a significantly higher stiffness than physically crosslinked gels [132, 149]. The elastic behavior of hydrogels can be described using the theory of rubber elasticity by Treloar and Flory [102] and later modified by Peppas and Merrill [117]. According to the theory of rubber elasticity, the elastic modulus of a crosslinked hydrogel network is [150].

$$E = \nu_E RT \tag{2}$$

Where ν_E, R and T are the density of elastically active chains, the gas constant, and absolute temperature, respectively. ν_E increases with increasing crosslink density or increasing the concentration of reactive groups in the hydrogel precursor solution. Therefore, one approach to increase hydrogel stiffness is to increase the concentration of reactive groups via increasing the macromer concentration in the gel precursor solution. For example, the compressive modulus of the SPEGA and SPELA gels increased from 5 kPa to 710 and 460 kPa, respectively, by increasing the reactive acrylate concentration from 0.02 to 0.13 mol/L (Fig. 5). The lower compressive modulus of SPELA compared to SPEGA can be attributed to the higher proximity of acrylates within the micelles in SPELA, leading to a higher degree of intra-molecular crosslinks [118, 151]. It is well established that intra-molecular crosslinks or loops are elastically inactive and don't contribute to the hydrogel stiffness [152].

5 Conclusions

The application of hydrogels in bioprinting as bio-ink and bio-paper is reviewed. The most important function of the hydrogel as the bio-ink is to mitigate the harsh conditions of bioprinting such as high temperature and shear stress on cells and to

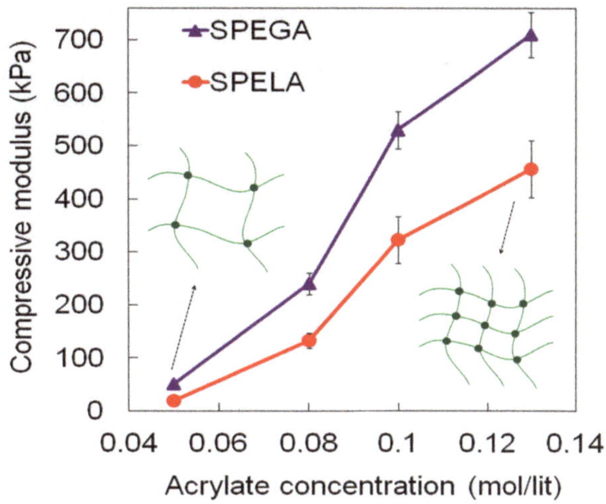

Fig. 5 Effect of reactive acrylate concentration on compressive modulus of *SPELA* and *SPEGA* hydrogels. The insets show the increase in crosslink density of the hydrogel concurrent with the increase in acrylate concentration

protect the encapsulated cells from the host immune system. Fast gelling alginate-calcium, fibrin-thrombin, and photo-activated polymerization are used as the bi-ink encapsulating the cells and printing the building block of engineered tissues. The bio-paper provides a temporary support for the cells printed with the bio-ink and it is removed after tissue printing. Hydrogels that are stimuli-sensitive, in situ crosslinkable, rapidly solidify after injection, and facilitate fusion of the printed cell aggregates are often used as the bio-paper. Important material properties in the selection of hydrogels for bioprinting include gelation kinetics, viscosity, viscoelastic response, permeability and transport properties, fusibility, and degradation kinetics. Ideally, the rate of degradation of the hydrogel as a bio-ink or bio-paper should match the rate of tissue formation and remodeling by the encapsulated cells. Future research should focus on developing ECM-mimetic hydrogels for drop-on-demand cell printing that satisfy the requirements of the target tissue.

Acknowledgements This work was supported by research grants to E. Jabbari from the National Science Foundation under grant Nos. DMR1049381, IIP-1357109, and CBET1403545, the National Institutes of Health under grant No. AR063745, and the AO Foundation under grant No. C10-44J. The authors thank Mr. Samuel Keeney (biomedical engineering, University of South Carolina) for editing the manuscript.

References

1. Song SJ, Choi J, Park YD, Lee JJ, Hong SY, Sun K. A three-dimensional bioprinting system for use with a hydrogel-based biomaterial and printing parameter characterization. Artif Organs. 2010;34:1044–8.
2. Calvert P. Printing cells. Science. 2007;318:208–9.
3. Guillotin B, Guillemot F. Cell patterning technologies for organotypic tissue fabrication. Trends Biotechnol. 2011;29:183–90.

4. Castel D, Pitaval A, Debily M-A, Gidrol X. Cell microarrays in drug discovery. Drug Discov Today. 2006;11:616–22.
5. Yarmush ML, King KR. Living-cell microarrays. Ann Rev Biomed Eng. 2009;11:235.
6. Ovsianikov A, Gruene M, Pflaum M, Koch L, Maiorana F, Wilhelmi M, Haverich A, Chichkov B. Laser printing of cells into 3D scaffolds. Biofabrication. 2010;2:014104.
7. Khalil S, Sun W. Bioprinting endothelial cells with alginate for 3d tissue constructs. J Biomech Eng. 2009;131:111002.
8. Malda J, Visser J, Melchels FP, Jungst T, Hennink WE, Dhert WJA, Groll J, Hutmacher DW. 25th anniversary article: engineering hydrogels for biofabrication. Adv Mater. 2013;25:5011–28.
9. Nakamura M, Iwanaga S, Henmi C, Arai K, Nishiyama Y. Biomatrices and biomaterials for future developments of bioprinting and biofabrication. Biofabrication. 2010;2:014110.
10. Koch L, Gruene M, Unger C, Chichkov B. Laser assisted cell printing. Curr Pharmaceut Biotechnol. 2013;14:91–7.
11. Calvert P. Inkjet printing for materials and devices. Chem Mater. 2001;13:3299–305.
12. Derby B. Printing and prototyping of tissues and scaffolds. Science. 2012;338:921–6.
13. Moon S, Hasan SK, Song YS, Xu F, Keles HO, Manzur F, Mikkilineni S, Hong JW, Nagatomi J, Haeggstrom E. Layer by layer three-dimensional tissue epitaxy by cell-laden hydrogel droplets. Tissue Eng Part C Methods. 2009;16:157–66.
14. Lee W, Pinckney J, Lee V, Lee J-H, Fischer K, Polio S, Park J-K, Yoo S-S. Three-dimensional bioprinting of rat embryonic neural cells. Neuroreport. 2009;20:798–803.
15. Murphy SV, Skardal A, Atala A. Evaluation of hydrogels for bio-printing applications. J Biomed Mater Res Part A. 2013;101:272–84.
16. Lee Y-B, Polio S, Lee W, Dai G, Menon L, Carroll RS, Yoo S-S. Bio-printing of collagen and VEGF-releasing fibrin gel scaffolds for neural stem cell culture. Exp Neurol. 2010;223:645–52.
17. Guillotin B, Souquet A, Catros S, Duocastella M, Pippenger B, Bellance S, Bareille R, Rémy M, Bordenave L, Amédée J. Laser assisted bioprinting of engineered tissue with high cell density and microscale organization. Biomaterials. 2010;31:7250–6.
18. Skardal A, Zhang J, McCoard L, Xu X, Oottamasathien S, Prestwich GD. Photocrosslinkable hyaluronan-gelatin hydrogels for two-step bioprinting. Tissue Eng Part A. 2010;16:2675–85.
19. Hong S, Song S-J, Lee JY, Jang H, Choi J, Sun K, Park Y. Cellular behavior in micropatterned hydrogels by bioprinting system depended on the cell types and cellular interaction. J Biosci Bioeng. 2013;116:224–30.
20. Skardal A, Zhang J, Prestwich GD. Bioprinting vessel-like constructs using hyaluronan hydrogels crosslinked with tetrahedral polyethylene glycol tetracrylates. Biomaterials. 2010;31:6173–81.
21. Pescosolido L, Schuurman W, Malda J, Matricardi P, Alhaique F, Coviello T, van Weeren PR, Dhert WJ, Hennink WE, Vermonden T. Hyaluronic acid and dextran-based semi-ipn hydrogels as biomaterials for bioprinting. Biomacromolecules. 2011;12:1831–8.
22. Bertassoni LE, Cardoso JC, Manoharan V, Cristino AL, Bhise NS, Araujo WA, Zorlutuna P, Vrana NE, Ghaemmaghami AM, Dokmeci MR. Direct-write bioprinting of cell-laden methacrylated gelatin hydrogels. Biofabrication. 2014;6:024105.
23. Mironov V, Prestwich G, Forgacs G. Bioprinting living structures. J Mater Chem. 2007;17:2054–60.
24. Buyukhatipoglu K, Chang R, Sun W, Clyne AM. Bioprinted nanoparticles for tissue engineering applications. Tissue Eng Part C Methods. 2009;16:631–42.
25. Schuurman W, Khristov V, Pot M, Van Weeren P, Dhert W, Malda J. Bioprinting of hybrid tissue constructs with tailorable mechanical properties. Biofabrication. 2011;3:021001.
26. Gasperini L, Maniglio D, Motta A, Migliaresi C. An electrohydrodynamic bioprinter for alginate hydrogels containing living cells. Tissue Eng Part C Methods. 2014;21:123–32.
27. Duan B, Hockaday LA, Kang KH, Butcher JT. 3D bioprinting of heterogeneous aortic valve conduits with alginate/gelatin hydrogels. J Biomed Mater Res A. 2013;101:1255–64.

28. Ilkhanizadeh S, Teixeira AI, Hermanson O. Inkjet printing of macromolecules on hydrogels to steer neural stem cell differentiation. Biomaterials. 2007;28:3936–43.
29. Kucukgul C, Ozler B, Karakas HE, Gozuacik D, Koc B. 3D hybrid bioprinting of macrovascular structures. Procedia Eng. 2013;59:183–92.
30. Hong SJ, Jin DP, Buck DW, Galiano RD, Mustoe TA. Impaired response of mature adipocytes of diabetic mice to hypoxia. Exp Cell Res. 2011;317:2299–307.
31. Fedorovich NE, Kuipers E, Gawlitta D, Dhert WJ, Alblas J. Scaffold porosity and oxygenation of printed hydrogel constructs affect functionality of embedded osteogenic progenitors. Tissue Eng Part A. 2011;17:2473–86.
32. Shim J-H, Lee J-S, Kim JY, Cho D-W. Bioprinting of a mechanically enhanced three-dimensional dual cell-laden construct for osteochondral tissue engineering using a multi-head tissue/organ building system. J Micromech Microeng. 2012;22:085014.
33. Norotte C, Marga FS, Niklason LE, Forgacs G. Scaffold-free vascular tissue engineering using bioprinting. Biomaterials. 2009;30:5910–7.
34. de Vos P, Faas MM, Strand B, Calafiore R. Alginate-based microcapsules for immunoisolation of pancreatic islets. Biomaterials. 2006;27:5603–17.
35. Murua A, Orive G, Hernández RM, Pedraz JL. Xenogeneic transplantation of erythropoietin-secreting cells immobilized in microcapsules using transient immunosuppression. J Control Release. 2009;137:174–8.
36. Chang PL, Van Raamsdonk JM, Hortelano G, Barsoum SC, MacDonald NC, Stockley TL. The *in vivo* delivery of heterologous proteins by microencapsulated recombinant cells. Trends Biotechnol. 1999;17:78–83.
37. Arai K, Iwanaga S, Toda H, Genci C, Nishiyama Y, Nakamura M. Three-dimensional inkjet biofabrication based on designed images. Biofabrication. 2011;3:034113.
38. Nishiyama Y, Nakamura M, Henmi C, Yamaguchi K, Mochizuki S, Nakagawa H, Takiura K. Development of a three-dimensional bioprinter: Construction of cell supporting structures using hydrogel and state-of-the-art inkjet technology. J Biomech Eng. 2009;131:035001.
39. Xu T, Olson J, Zhao W, Atala A, Zhu J-M, Yoo JJ. Characterization of cell constructs generated with inkjet printing technology using *in vivo* magnetic resonance imaging. J Manuf Sci Eng. 2008;130:021013.
40. Xu T, Gregory CA, Molnar P, Cui X, Jalota S, Bhaduri SB, Boland T. Viability and electrophysiology of neural cell structures generated by the inkjet printing method. Biomaterials. 2006;27:3580–8.
41. Cui X, Boland T. Human microvasculature fabrication using thermal inkjet printing technology. Biomaterials. 2009;30:6221–7.
42. Cui X, Breitenkamp K, Finn M, Lotz M, D'Lima DD. Direct human cartilage repair using three-dimensional bioprinting technology. Tissue Eng Part A. 2012;18:1304–12.
43. Landers R, Pfister A, Hübner U, John H, Schmelzeisen R, Mülhaupt R. Fabrication of soft tissue engineering scaffolds by means of rapid prototyping techniques. J Mater Sci. 2002;37:3107–16.
44. Mann BK, Gobin AS, Tsai AT, Schmedlen RH, West JL. Smooth muscle cell growth in photopolymerized hydrogels with cell adhesive and proteolytically degradable domains: synthetic ECM analogs for tissue engineering. Biomaterials. 2001;22:3045–51.
45. VandeVondele S, Vörös J, Hubbell JA. RGD-grafted poly-L-lysine-graft-(polyethylene glycol) copolymers block non-specific protein adsorption while promoting cell adhesion. Biotechnol Bioeng. 2003;82:784–90.
46. Huang N-P, Vörös J, De Paul SM, Textor M, Spencer ND. Biotin-derivatized poly (L-lysine)-g-poly (ethylene glycol): a novel polymeric interface for bioaffinity sensing. Langmuir. 2002;18:220–30.
47. Yang F, Williams CG, D-a W, Lee H, Manson PN, Elisseeff J. The effect of incorporating rgd adhesive peptide in polyethylene glycol diacrylate hydrogel on osteogenesis of bone marrow stromal cells. Biomaterials. 2005;26:5991–8.

48. Elisseeff J, McIntosh W, Anseth K, Riley S, Ragan P, Langer R. Photoencapsulation of chondrocytes in poly (ethylene oxide)-based semi-interpenetrating networks. J Biomed Mater Res. 2000;51:164–71.
49. Jeon O, Bouhadir KH, Mansour JM, Alsberg E. Photocrosslinked alginate hydrogels with tunable biodegradation rates and mechanical properties. Biomaterials. 2009;30:2724–34.
50. Fedorovich NE, Swennen I, Girones J, Moroni L, Van Blitterswijk CA, Schacht E, Alblas J, Dhert WJ. Evaluation of photocrosslinked lutrol hydrogel for tissue printing applications. Biomacromolecules. 2009;10:1689–96.
51. Morota K, Matsumoto H, Mizukoshi T, Konosu Y, Minagawa M, Tanioka A, Yamagata Y, Inoue K. Poly (ethylene oxide) thin films produced by electrospray deposition: morphology control and additive effects of alcohols on nanostructure. J Colloid Interf Sci. 2004;279:484–92.
52. Cruise GM, Hegre OD, Scharp DS, Hubbell JA. A sensitivity study of the key parameters in the interfacial photopolymerization of poly (ethylene glycol) diacrylate upon porcine islets. Biotechnol Bioeng. 1998;57:655–65.
53. Ferris CJ, Gilmore KJ, Beirne S, McCallum D, Wallace GG, het Panhuis M. Bio-ink for on-demand printing of living cells. Biomater Sci. 2013;1:224–30.
54. Moon S, Ceyhan E, Gurkan UA, Demirci U. Statistical modeling of single target cell encapsulation. PLoS One. 2011;6:e21580.
55. Jakab K, Neagu A, Mironov V, Markwald RR, Forgacs G. Engineering biological structures of prescribed shape using self-assembling multicellular systems. Proc Natl Acad Sci U S A. 2004;101:2864–9.
56. Jeong B, Lee KM, Gutowska A, An YH. Thermogelling biodegradable copolymer aqueous solutions for injectable protein delivery and tissue engineering. Biomacromolecules. 2002;3:865–8.
57. Ohya S, Nakayama Y, Matsuda T. Thermoresponsive artificial extracellular matrix for tissue engineering: hyaluronic acid bioconjugated with poly (n-isopropylacrylamide) grafts. Biomacromolecules. 2001;2:856–63.
58. Molinaro G, Leroux J-C, Damas J, Adam A. Biocompatibility of thermosensitive chitosan-based hydrogels: an *in vivo* experimental approach to injectable biomaterials. Biomaterials. 2002;23:2717–22.
59. Iwami K, Noda T, Ishida K, Morishima K, Nakamura M, Umeda N. Bio rapid prototyping by extruding/aspirating/refilling thermoreversible hydrogel. Biofabrication. 2010;2:014108.
60. Landers R, Hübner U, Schmelzeisen R, Mülhaupt R. Rapid prototyping of scaffolds derived from thermoreversible hydrogels and tailored for applications in tissue engineering. Biomaterials. 2002;23:4437–47.
61. Biase M D, Saunders RE, Tirelli N, Derby B. Inkjet printing and cell seeding thermoreversible photocurable gel structures. Soft Matter. 2011;7:2639–46.
62. Boland T, Mironov V, Gutowska A, Roth E, Markwald RR. Cell and organ printing 2: fusion of cell aggregates in three-dimensional gels. Anat Rec A Discov Mol Cell Evol Biol. 2003;272:497–502.
63. Elisseeff J, Anseth K, Sims D, McIntosh W, Randolph M, Yaremchuk M, Langer R. Transdermal photopolymerization of poly (ethylene oxide)-based injectable hydrogels for tissue-engineered cartilage. Plast Reconstr Surg. 1999;104:1014–22.
64. Baier Leach J, Bivens KA, Patrick CW Jr, Schmidt CE. Photocrosslinked hyaluronic acid hydrogels: natural, biodegradable tissue engineering scaffolds. Biotechnol Bioeng. 2003;82:578–89.
65. Park YD, Tirelli N, Hubbell JA. Photopolymerized hyaluronic acid-based hydrogels and interpenetrating networks. Biomaterials. 2003;24:893–900.
66. Tsang VL, Chen AA, Cho LM, Jadin KD, Sah RL, DeLong S, West JL, Bhatia SN. Fabrication of 3D hepatic tissues by additive photopatterning of cellular hydrogels. FASEB J. 2007;21:790–801.
67. Gyenes T, Torma V, Gyarmati B, Zrínyi M. Synthesis and swelling properties of novel pH-sensitive poly (aspartic acid) gels. Acta Biomater. 2008;4:733–44.

68. Lutolf M, Lauer-Fields J, Schmoekel H, Metters A, Weber F, Fields G, Hubbell J. Synthetic matrix metalloproteinase-sensitive hydrogels for the conduction of tissue regeneration: engineering cell-invasion characteristics. Proc Natl Acad Sci U S A. 2003;100:5413–8.
69. Mironov V, Kasyanov V, Markwald R. Bioprinting: Directed tissue self-assembly. Chem Eng Prog. 2007;103:S12.
70. Mehra TD, Ghosh K, Shu XZ, Prestwich GD, Clark RA. Molecular stenting with a crosslinked hyaluronan derivative inhibits collagen gel contraction. J Invest Derm. 2006;126:2202–9.
71. Moon JJ, Hahn MS, Kim I, Nsiah BA, West JL. Micropatterning of poly (ethylene glycol) diacrylate hydrogels with biomolecules to regulate and guide endothelial morphogenesis. Tissue Eng Part A. 2008;15:579–85.
72. Wylie RG, Ahsan S, Aizawa Y, Maxwell KL, Morshead CM, Shoichet MS. Spatially controlled simultaneous patterning of multiple growth factors in three-dimensional hydrogels. Nat Mater. 2011;10:799–806.
73. Moon JJ, West JL. Vascularization of engineered tissues: Approaches to promote angio-genesis in biomaterials. Curr Topics Med Chem. 2008;8:300–10.
74. Leslie-Barbick JE, Shen C, Chen C, West JL. Micron-scale spatially patterned, covalently immobilized vascular endothelial growth factor on hydrogels accelerates endothelial tubulogenesis and increases cellular angiogenic responses. Tissue Eng Part A. 2010;17:221–9.
75. Saik JE, Gould DJ, Watkins EM, Dickinson ME, West JL. Covalently immobilized platelet-derived growth factor-BB promotes angiogenesis in biomimetic poly (ethylene glycol) hydrogels. Acta Biomater. 2011;7:133–43.
76. Wylie RG, Shoichet MS. Three-dimensional spatial patterning of proteins in hydrogels. Biomacromolecules. 2011;12:3789–96.
77. Palecek SP, Loftus JC, Ginsberg MH, Lauffenburger DA, Horwitz AF. Integrin-ligand binding properties govern cell migration speed through cell-substratum adhesiveness. Nature. 1997;385:537–40.
78. Nuttelman CR, Mortisen DJ, Henry SM, Anseth KS. Attachment of fibronectin to poly (vinyl alcohol) hydrogels promotes NIH3T3 cell adhesion, proliferation, and migration. J Biomed Mater Res. 2001;57:217–23.
79. Benoit DS, Anseth KS. Heparin functionalized PEG gels that modulate protein adsorption for hMSC adhesion and differentiation. Acta Biomater. 2005;1:461–70.
80. Geckil H, Xu F, Zhang X, Moon S, Demirci U. Engineering hydrogels as extracellular matrix mimics. Nanomedicine. 2010;5:469–84.
81. Prestwich GD, Shu XZ, Liu Y, Cai S, Walsh JF, Hughes CW, et al. Injectable synthetic extracellular matrices for tissue engineering and repair. In: Fisher JP, Editor. Tissue Engineering. New York: Springer; 2007. pp. 125–33.
82. Neagu A, Jakab K, Jamison R, Forgacs G. Role of physical mechanisms in biological self-organization. Phys Rev Lett. 2005;95:178104.
83. Pérez-Pomares JM, Foty RA. Tissue fusion and cell sorting in embryonic development and disease: biomedical implications. Bioessays. 2006;28:809–21.
84. Jakab K, Norotte C, Damon B, Marga F, Neagu A, Besch-Williford CL, Kachurin A, Church KH, Park H, Mironov V. Tissue engineering by self-assembly of cells printed into topologically defined structures. Tissue Eng Part A. 2008;14:413–21.
85. Forgacs G, Foty RA, Shafrir Y, Steinberg MS. Viscoelastic properties of living embryonic tissues: a quantitative study. Biophys J. 1998;74:2227–34.
86. Foty RA, Steinberg MS. The differential adhesion hypothesis: a direct evaluation. Dev Biol. 2005;278:255–63.
87. Robinson EE, Zazzali KM, Corbett SA, Foty RA. A5β1 integrin mediates strong tissue cohesion. J Cell Sci. 2003;116:377–86.
88. Mironov V, Boland T, Trusk T, Forgacs G, Markwald RR. Organ printing: computer-aided jet-based 3D tissue engineering. Trends Biotechnol. 2003;21:157–61.
89. Rago AP, Dean DM, Morgan JR. Controlling cell position in complex heterotypic 3d micro-tissues by tissue fusion. Biotechnol Bioeng. 2009;102:1231–41.

90. Mironov V, Visconti RP, Kasyanov V, Forgacs G, Drake CJ, Markwald RR. Organ printing: tissue spheroids as building blocks. Biomaterials. 2009;30:2164–74.
91. Roth EA, Xu T, Das M, Gregory C, Hickman JJ, Boland T. Inkjet printing for high-throughput cell patterning. Biomaterials. 2004;25:3707–15.
92. Wilson WC, Boland T. Cell and organ printing 1: protein and cell printers. Anat Rec A Discov Mol Cell Evol Biol. 2003;272:491–6.
93. Jakab K, Damon B, Neagu A, Kachurin A, Forgacs G. Three-dimensional tissue constructs built by bioprinting. Biorheology. 2006;43:509–13.
94. Xu T, Jin J, Gregory C, Hickman JJ, Boland T. Inkjet printing of viable mammalian cells. Biomaterials. 2005;26:93–9.
95. Boland T, Tao X, Damon BJ, Manley B, Kesari P, Jalota S, Bhaduri S. Drop-on-demand printing of cells and materials for designer tissue constructs. Mater Sci Eng C. 2007;27:372–6.
96. Shu XZ, Ahmad S, Liu Y, Prestwich GD. Synthesis and evaluation of injectable, in situ crosslinkable synthetic extracellular matrices for tissue engineering. J Biomed Mater Res Part A. 2006;79:902–12.
97. Shu XZ, Liu Y, Palumbo F, Prestwich GD. Disulfide-crosslinked hyaluronan-gelatin hydrogel films: a covalent mimic of the extracellular matrix for *in vitro* cell growth. Biomaterials. 2003;24:3825–34.
98. Shu XZ, Ghosh K, Liu Y, Palumbo FS, Luo Y, Clark RA, Prestwich GD. Attachment and spreading of fibroblasts on an RGD peptide-modified injectable hyaluronan hydrogel. J Biomed Mater Res Part A. 2004;68:365–75.
99. Ghosh K, Ren X-D, Shu XZ, Prestwich GD, Clark RA. Fibronectin functional domains coupled to hyaluronan stimulate adult human dermal fibroblast responses critical for wound healing. Tissue Eng. 2006;12:601–13.
100. Schuurman W, Levett PA, Pot MW, van Weeren PR, Dhert WJA, Hutmacher DW, Melchels FPW, Klein TJ, Malda J. Gelatin-methacrylamide hydrogels as potential biomaterials for fabrication of tissue-engineered cartilage constructs. Macromol Biosci. 2013;13:551–61.
101. Dababneh AB, Ozbolat IT. Bioprinting technology: a current state-of-the-art review. J Manuf Sci Eng Trans ASME. 2014;136:061016.
102. Flory PJ. Principles of polymer chemistry. New York: Cornell University Press; 1953.
103. Cushing MC, Anseth KS. Hydrogel cell cultures. Science. 2007;316:1133–4.
104. Moeinzadeh S, Barati D, He XZ, Jabbari E. Gelation characteristics and osteogenic differentiation of stromal cells in inert hydrolytically degradable micellar polyethylene glycol hydrogels. Biomacromolecules. 2012;13:2073–86.
105. Moeinzadeh S, Jabbari E. Nanostructure formation in hydrogels. In: Bhushan B et al. editors. Handbook of nanomaterials properties. Berlin: Springer Berlin Heidelberg; 2014. pp. 285–97.
106. Cohen DL, Malone E, Lipson H, Bonassar LJ. Direct freeform fabrication of seeded hydrogels in arbitrary geometries. Tissue Eng. 2006;12:1325–35.
107. O'Lenick TG, Jin NX, Woodcock JW, Zhao B. Rheological properties of aqueous micellar gels of a thermo- and pH-sensitive aABA triblock copolymer. J Phys Chem B. 2011;115:2870–81.
108. Moeinzadeh S, Barati D, Sarvestani SK, Karaman O, Jabbari E. Nanostructure formation and transition from surface to bulk degradation in polyethylene glycol gels chain-extended with short hydroxy acid segments. Biomacromolecules. 2013;14:2917–28.
109. Moeinzadeh S, Jabbari E. Mesoscale simulation of the effect of a lactide segment on the nanostructure of star poly(ethylene glycol-co-lactide)-acrylate macromonomers in aqueous solution. J Phys Chem B. 2012;116:1536–43.
110. Moeinzadeh S, Khorasani SN, Ma JY, He X, Jabbari E. Synthesis and gelation characteristics of photo-crosslinkable star poly (ethylene oxide-co-lactide-glycolide acrylate) macromonomers. Polymer. 2011;52:3887–96.
111. Odian G. Principles of polymerization. New York: Wiley; 1981.

112. Queen D, Gaylor JDS, Evans JH, Courtney JM, Reid WH. The preclinical evaluation of the water-vapor transmission rate through burn wound dressings. Biomaterials. 1987;8:367–71.
113. Amsden B. Solute diffusion within hydrogels. Mechanisms and models. Macromolecules. 1998;31:8382–95.
114. Peppas NA, Lustig SR. Solute diffusion in hydrophilic network structures. In: Peppas NA, editor. Hydrogels in medicine and pharmacy Vol. I Fundamentals. Boca Raton: CRC Press; 1986. pp. 57–84.
115. Ganji F, Vasheghani-Farahani S, Vasheghani-Farahani E. Theoretical description of hydrogel swelling: a review. Iranian Polym J. 2010;19:375–98.
116. Sukumar VS, Lopina ST. Network model for the swelling properties of end-linked linear and star poly(ethylene oxide) hydrogels. Macromolecules. 2002;35:10189–92.
117. Peppas NA, Bures P, Leobandung W, Ichikawa H. Hydrogels in pharmaceutical formulations. Eur J Pharmaceut Biopharmaceut. 2000;50:27–46.
118. Barati D, Moeinzadeh S, Karaman O, Jabbari E. Time dependence of material properties of polyethylene glycol hydrogels chain extended with short hydroxy acid segments. Polymer. 2014;55:3894–904.
119. Khetan S, Guvendiren M, Legant WR, Cohen DM, Chen CS, Burdick JA. Degradation-mediated cellular traction directs stem cell fate in covalently crosslinked three-dimensional hydrogels. Nat Mater. 2013;12:458–65.
120. Chatterjee K, Lin-Gibson S, Wallace WE, Parekh SH, Lee YJ, Cicerone MT, Young MF, Simon CG. The effect of 3D hydrogel scaffold modulus on osteoblast differentiation and mineralization revealed by combinatorial screening. Biomaterials. 2010;31:5051–62.
121. Chen YC, Lin RZ, Qi H, Yang YZ, Bae HJ, Melero-Martin JM, Khademhosseini A. Functional human vascular network generated in photocrosslinkable gelatin methacrylate hydrogels. Adv Funct Mater. 2012;22:2027–39.
122. Henderson JA, He X, Jabbari E. Concurrent differentiation of marrow stromal cells to osteogenic and vasculogenic lineages. Macromol Biosci. 2008;8:499–507.
123. Ozbolat IT, Yu Y. Bioprinting toward organ fabrication: Challenges and future trends. IEEE Trans Biomed Eng. 2013;60:691–9.
124. Jin R, Teixeira LSM, Krouwels A, Dijkstra PJ, van Blitterswijk CA, Karperien M, Feijen J. Synthesis and characterization of hyaluronic acid-poly(ethylene glycol) hydrogels via michael addition: an injectable biomaterial for cartilage repair. Acta Biomater. 2010;6:1968–77.
125. Ye Q, Zund G, Benedikt P, Jockenhoevel S, Hoerstrup SP, Sakyama S, Hubbell JA, Turina M. Fibrin gel as a three dimensional matrix in cardiovascular tissue engineering. Eur J Cardio Thor Surg. 2000;17:587–91.
126. Ashton RS, Banerjee A, Punyani S, Schaffer DV, Kane RS. Scaffolds based on degradable alginate hydrogels and poly(lactide-co-glycolide) microspheres for stem cell culture. Biomaterials. 2007;28:5518–25.
127. Kharkar PM, Kiick KL, Kloxin AM. Designing degradable hydrogels for orthogonal control of cell microenvironments. Chem Soc Rev. 2013;42:7335–72.
128. Lutolf MR, Weber FE, Schmoekel HG, Schense JC, Kohler T, Muller R, Hubbell JA. Repair of bone defects using synthetic mimetics of collagenous extracellular matrices. Nat Biotechnol. 2003;21:513–8.
129. Kraehenbuehl TP, Ferreira LS, Zammaretti P, Hubbell JA, Langer R. Cell-responsive hydrogel for encapsulation of vascular cells. Biomaterials. 2009;30:4318–24.
130. Ko CY, Yang CY, Yang SR, Ku KL, Tsao CK, Chuang DCC, Chu IM, Cheng MH. Cartilage formation through alterations of amphiphilicity of poly(ethylene glycol)-poly(caprolactone) copolymer hydrogels. RSC Adv. 2013;3:25769–79.
131. Nakayama Y, Okuda K, Takamizawa K, Nakayama A. Preparation of well-defined poly(ether-ester) macromers: photogelation and biodegradability. Acta Biomater. 2011;7:1496–503.

132. Sarvestani AS, Xu W, He X, Jabbari E. Gelation and degradation characteristics of in situ photo-crosslinked poly (l-lactide-co-ethylene oxide-co-fumarate) hydrogels. Polymer. 2007;48:7113–20.
133. Moeinzadeh S, Khorasani SN, Ma J, He X, Jabbari E. Synthesis and gelation characteristics of photo-crosslinkable star poly (ethylene oxide-co-lactide-glycolide acrylate) macromonomers. Polymer. 2011;52:3887–96.
134. He JL, Zhang MZ, Ni PH. Rapidly in situ forming polyphosphoester-based hydrogels for injectable drug delivery carriers. Soft Matter. 2012;8:6033–8.
135. Wang YC, Lee WJ, Ju SP. Modeling of the polyethylene and poly(L-lactide) triblock copolymer: a dissipative particle dynamics study. J Chem Phys. 2009;131:124901.
136. Rodriguez-Galan A, Franco L, Puiggali J. Degradable poly(ester amide)s for biomedical applications. Polymers. 2011;3:65–99.
137. Zheng YT, He CL, Huynh CT, Lee DS. Biodegradable ph- and temperature-sensitive multiblock copolymer hydrogels based on poly(amino-ester urethane)s. Macromol Res. 2010;18:974–80.
138. Zhang NL, Kohn DH. Using polymeric materials to control stem cell behavior for tissue regeneration. Birth Defects Res C Embryo Today Rev. 2012;96:63–81.
139. Drury JL, Dennis RG, Mooney DJ. The tensile properties of alginate hydrogels. Biomaterials. 2004;25:3187–99.
140. Koob TJ, Hernandez DJ. Mechanical and thermal properties of novel polymerized ndga-gelatin hydrogels. Biomaterials. 2003;24:1285–92.
141. Gu W, Yao H, Huang C, Cheung H. New insight into deformation-dependent hydraulic permeability of gels and cartilage, and dynamic behavior of agarose gels in confined compression. J Biomech. 2003;36:593–8.
142. Ahearne M, Yang Y, Haj AJ E, Then KY, Liu K-K. Characterizing the viscoelastic properties of thin hydrogel-based constructs for tissue engineering applications. J R Soc Interface. 2005;2:455–63.
143. Constantinides G, Kalcioglu ZI, McFarland M, Smith JF, Van Vliet KJ. Probing mechanical properties of fully hydrated gels and biological tissues. J Biomech. 2008;41:3285–9.
144. Gabler S, Stampfl J, Koch T, Seidler S, Schuller G, Redl H, Juras V, Trattnig S, Weidisch R. Determination of the viscoelastic properties of hydrogels based on polyethylene glycol diacrylate (peg-da) and human articular cartilage. Int J Mater Eng Innov. 2009;1:3–20.
145. Huebsch N, Arany PR, Mao AS, Shvartsman D, Ali OA, Bencherif SA, Rivera-Feliciano J, Mooney DJ. Harnessing traction-mediated manipulation of the cell/matrix interface to control stem-cell fate. Nat Mater. 2010;9:518–26.
146. Aamer KA, Sardinha H, Bhatia SR, Tew GN. Rheological studies of PLLA-PEO-PLLA triblock copolymer hydrogels. Biomaterials. 2004;25:1087–93.
147. Pollock JF, Healy KE. Mechanical and swelling characterization of poly(n-isopropyl acrylamide-co-methoxy poly(ethylene glycol) methacrylate) sol-gels. Acta Biomater. 2010;6:1307–18.
148. Nichol JW, Koshy ST, Bae H, Hwang CM, Yamanlar S, Khademhosseini A. Cell-laden microengineered gelatin methacrylate hydrogels. Biomaterials. 2010;31:5536–44.
149. Amsden BG, Misra G, Gu F, Younes HM. Synthesis and characterization of a photo-cross-linked biodegradable elastomer. Biomacromolecules. 2004;5:2479–86.
150. Cima LG, Lopina ST. Network structures of radiation-crosslinked star polymer gels. Macromolecules. 1995;28:6787–94.
151. Elliott JE, Bowman CN. Kinetics of primary cyclization reactions in cross-linked polymers: an analytical and numerical approach to heterogeneity in network formation. Macromolecules. 1999;32:8621–8.
152. Sarvestani AS, Xu WJ, He X, Jabbari E. Gelation and degradation characteristics of in situ photo-crosslinked poly(l-lactid-co-ethylene oxide-co-fumarate) hydrogels. Polymer. 2007;48:7113–20.

Three-Dimensional Bioprinting in Regenerative Medicine

Xiaofeng Cui

1 Introduction

In 1993, Langer and Vacanti first defined tissue engineering as an approach of seeding cells to the pre-formed solid and rigid biomaterial scaffolds for tissue fabrication [1]. However, the term of tissue engineering was introduced even earlier by Dr. Fung of the University of California at San Diego in 1985 [2]. In conventional tissue engineering approach, the autologous cells are first cultured in monolayer to expand the cell numbers. The cultured cells are then collected and seeded into the preformed porous scaffolds. The scaffolds used for tissue engineering should be biocompatible and degradable. The seeded cells on the scaffold are kept alive and can penetrate or migrate inside the scaffolds instead of staying on the surface. Therefore, the tissue engineering scaffolds should be highly porous with inter-connected pores and safe to the seeded cells. In addition, a customized bioreactor mimicking *in vivo* environment and stimulation is usually desired to maturate the fabricated organ construct before implantation. The goal of tissue engineering is to create the replacements for the lost or diseased organs and eventually solve the crisis of organ donor shortage. Some successes have been achieved in engineering thin and hollow organs [3, 4]. These tissues can survive *in vivo* through nutrients diffusion from the

X. Cui (✉)
School of Chemistry, Chemical Engineering and Life Sciences, Wuhan University of Technology, 122 Luoshi Road, Wuhan, Hubei, China
e-mail: xfc.cui@gmail.com

Center for Biotechnology and Interdisciplinary Studies, Rensselaer Polytechnic Institute, 110 Eighth St, Troy, NY 12180, USA

Stemorgan Therapeutics, Albany, NY 12207, USA

K. Turksen (ed.), *Bioprinting in Regenerative Medicine*, Stem Cell Biology and Regenerative Medicine, DOI 10.1007/978-3-319-21386-6_5

host vasculature. However, more than 90% demanding organs are thick and complex, such as kidney, liver, and heart (OPTN & SRTR Annual Data Report 2010). When the size of engineered tissue exceeds 400 μm in any dimension, it will surpass the oxygen diffusion limitation. In this case, functional vasculature must be enabled in the engineered constructs to supply the cells with oxygen and nutrients, and also to remove the waste products generated by the tissue [5]. Unfortunately, the conventional tissue engineering approaches failed to generate these thick, complex and vascularized tissues due to these limitations:

a. The effectiveness of cell seeding and penetration to the biomaterial scaffold is still limited. Although scaffold design has been significantly improved to enhance the cell seeding and migration, the uniform of tissue formation or maturation throughout the scaffold is still far from optimal [6–8].
b. Organs with complex structure are usually composed by multiple cell types and biological factors. However, the precise delivery of cells and biological factors to the desired 3D positions is still far from being resolved.
c. Thick tissues possess complex vascular system [9], which should be enabled within the scaffold. However, the conventional tissue engineering approach has difficulties to construct vascular system within the pre-formed 3D scaffolds.

Additive manufacturing or 3D printing is driving significant innovations in manufacturing, engineering, education and medicine. 3D bioprinting, which was derived by combing biotechnology and 3D printing, is promising to solve these critical issues mentioned above. As one of the most advanced enabling technology in tissue engineering, 3D bioprinting combines solid freeform fabrication and precise placement of cells and other biological factors to the desired 2D and 3D positions. It is described as a precise approach of delivering biomaterials, cells and supporting biological factors to the targeted locations with spatial control to fabricate functional 3D constructs. The key elements of realizing functional bioprinting include capacity of precise positioning, printable biomaterials, and cell sources. In addition, vascularization, innervation, and maturation are also crucial to engineer functional tissues. Bioprinting has promising applications in the field of regenerative medicine, personalized medicine, clinical diagnosis and medicinal development. Although the concept of bioprinting was introduced more than 10 years ago, the current progress of bioprinting is still in its initial stage and far from industrial applications.

The three most common bioprinting mechanisms are inkjet bioprinting [10–21], extrusion bioprinting [22–24], and laser bioprinting [25–27]. Extrusion bioprinting is a contact printing process and typically uses temperature-controlled polymerized materials for scaffold fabrication. This printing process usually causes high cell casualty so it is frequently used in acellular material printing. Sometimes extrusion bioprinting also applies in cell spheroids deposition. This approach does not demand high printing resolution and it is more likely a dispensing process instead of printing. In addition, this approach has difficulties of managing singe cell which is critically important for neuron regeneration or fabricating functional tissues with higher degree of cell organization of specific anatomic structures [28, 29]. Laser bioprinting offers higher cell viability and printing resolution. Instead of moving cells directly, laser bioprinting uses laser energy to vaporize the solution of biological samples and eject the remaining substances [25]. This approach may cause over-drying leading to the failure for biological systems. Furthermore, the much higher cost of laser printing equipment, as well as the exceedingly low printing efficiency inhibit its application in regenerative medicine [30, 31]. Thus it is mostly applied in the basic research field when single or multiple cell manipulation is needed, instead of tissue construction or other clinical applications demanding higher throughput.

Inkjet printing is also known as drop-on-demand printing. It is a non-contact printing technology that reproduces digital patterns onto a substrate using tiny ink drops [32]. Inkjet printing is based on thermal, piezoelectric, or electromagnetic mechanisms [33]. In thermal inkjet printers, small air bubbles generated by heating in the printhead collapse to provide pressure pulses to eject tiny drops out of the nozzle [34–36]. The droplet size varies from 10 to 150 pL, which is determined by the applied temperature gradient, frequency of current pulses, and viscosity of the ink [34–36]. As for the piezoelectric inkjet printers, the actuator of polycrystalline piezoelectric ceramic in each nozzle provides the transient pressure to eject the ink drops [37]. These printing technologies have already been widely used in printing electronic materials and complex integrated circuits in industry [38]. Although biological substances are usually considered sensitive, fragile DNA molecules have been directly printed using commercially available inkjet printers for high-density DNA microarray fabrication [39, 40]. Challenges still exist when printing cells using inkjet technology. The working frequency of piezoelectric inkjet printers is 15–25 kHz, which is within the well-documented sonification damage to the cell membrane [41]. Although the heating element in thermal inkjet printers raises the local temperature to 300 °C [36], the ejected mammalian cells are only heated for 2 μs with a temperature raise of 4–10 °C above ambient and an average cell viability over 90 % [11]. In addition, the development, operation, and maintenance of thermal inkjet is usually more convenient than piezoelectric printing. Therefore, the majority successes in tissue bioprinting are based on thermal inkjet printing instead of piezoelectric inkjet printing. One limitation of inkjet bioprinting is the strict requirement of bioink viscosity. This issue has recently been minimized by using water based biomaterials or combination of various printing technologies. Water based bioink allows the printer to freely deliver cells from single to multiple cells by simply adjusting the bioink concentration and the digital patterns. Cells are usually well-protected in the aqueous environment during the printing process therefore it is assumed to be the safest strategy to deliver living systems.

Based on the discussion above, bioprinting based on thermal inkjet printing is so far the most appropriate approach for regenerative medicine and tissue engineering applications. Researchers keep developing this technology as an optimal approach for cell delivery and scaffold fabrication. Therefore, we will mainly focus on the advancement and applications using this bioprinting technology in this chapter.

2 Cell Printing

Although the term of bioprinting can be used on printing any biological systems, it usually involves living cell patterning in tissue engineering and regenerative medicine applications. Therefore, the capacity of printing living cells is critical to evaluate a bioprinting platform or system.

Although bioprinting based on thermal inkjet printing technology has many successful applications, there were concerns that the printing process may cause damages or cell death. The small printhead nozzle size is necessary for high printing resolution. Due to the thermal heat and mechanical stresses applied to the cells during printing, it is possible that the cells may be damaged or their phenotype may be altered [42]. Therefore, a comprehensive evaluation of cell viability, apoptosis, heat shock proteins production, cell membrane damages of the printed cells is desired to confirm the bioprinting safety. Using a modified Hewlett-Packard (HP) thermal inkjet printer, cell viability at various cell concentrations was between 85 and 95 %. No significant difference in apoptosis and heat shock protein expression was observed between printed and non-printed cells [11]. Quantitative cell seeding can be achieved by adjusting the cell concentration in bioink. The inkjet printing process does alter the cell membrane of printed cells. Fluorescent labeled dextran dye with molecular weight (MW) up to 40,000 can penetrate into the printed cells. No dye was found in the non-printed cells even with the lowest MW (3000) (Fig. 1). The cell membrane pore size was estimated as 105 Å according to the Stokes diameter of these dye molecules [11].

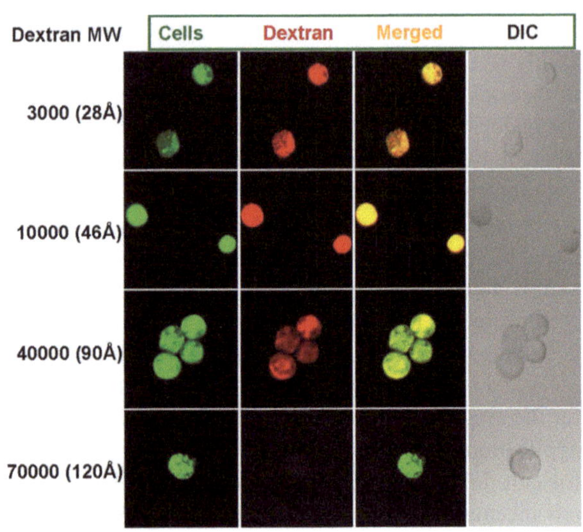

Fig. 1 Printed cell membrane pore evaluation using dextran dye penetration study

The pores developed during printing were transient and can be repaired by the cells in just a couple hours. The transient nature of the cell membrane pores as well as the self-repair mechanism can be utilized for targeted gene delivery during the printing process [11, 43].

3 Microvasculature Printing

Although the concept of tissue engineering was introduced more than two decades ago, the current tissue engineering strategies still cannot create fully vascularized tissue constructs. The current tissue engineering paradigm is that successfully engineered thick tissues must include vasculature. As biological approaches alone, such as VEGF or co-culture of vessel cells, have fallen short of their promises, one may look for an engineering approach to build microvasculature. Layer-by-layer approaches for customized fabrication of cell and scaffold constructs have shown great potential in building complex 3D structures [44]. With the advent of cell printing, one may be able to build precise human microvasculature with suitable bioink. Human microvascular endothelial cells (HMVECs) and fibrin scaffold were utilized as bioink for microvasculature construction [12].

A standard inkjet printer was modified to simultaneously deposit HMVECs and fibrin scaffold to form the microvasculature. The bioink and biopaper components for fibrin bioprinting were carefully evaluated for optimal condition of simultaneous deposition of cells and scaffold [12]. The printed microvasculature was incubated for 10–15 min after the printing to finalize the crosslinking and enhance the cell attachment.

After 3 weeks in culture, the printed HMVECs aligned themselves in the fibrin channel and proliferated to form a confluent lining. Confocal laser scanning images at the z-axis demonstrated tubular structure of the printed human microvasculature. The endothelial cells were forming a vessel-like structure in the printed fibrin channel [12]. This demonstrates the printed and proliferated endothelial cells possessed the crucial angiogenesis function. The simultaneous deposition of endothelial cells and fibrin using thermal inkjet printing technology can be used for human microvasculature fabrication (Fig. 2).

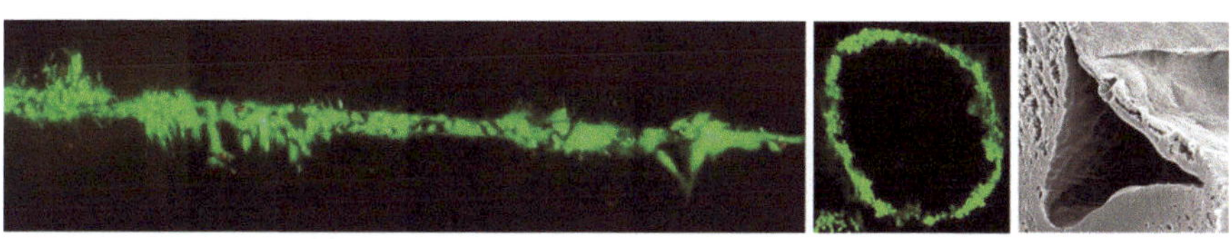

Fig. 2 Inkjet bioprinted human microvasculature using HMVEC and fibrin

4 Muscle Printing

Biological microelectromechanical system (Bio-MEMS) devices conjugated with biological components are promising for the development of novel bioengineering microdevices, such as motors and actuators [45], heart pumps [46], and biosensors [47]. Muscle cells have been widely used in these applications by generating force activated by actin-myosin motors regulated by excitation-contraction coupling [48]. These muscle powered microdevices utilizing energy generated by biochemical reaction are promising to save energy, resources, and spaces [49]. C2C12 skeletal muscle cells possess the advantages of infinite proliferation and differentiation into multinucleated myotubes [50]. As a well established cell line, the overall properties of C2C12 cells cultured and differentiated *in vitro* have been tested to closely mimic the properties of skeletal muscle *in vivo* [51]. Although C2C12 cells have been widely used to incorporate with bio-microdevices for many applications, it is important that the muscle cells and microdevices are consistently conjugated to produce reliable and reproducible results. The traditional methodology for Bio-MEMS fabrication is to manually seed cells on or into the microdevices [52]. However, the randomly deposited cells through this approach were uneven and further affected the cell proliferation and differentiation. Therefore, it is critical to incorporate a precise cell seeding technology to develop the Bio-MEMS constructs with consistent cell arrangement.

Bioprinting was able to print and align C2C12 cells onto the tiny cantilevers at a resolution at 300 dpi (85 μm). In order to control the cell proliferation and differentiation with minimal variations, same amount of cells were printed to evenly cover each cantilever of the microdevices. The viability of printed C2C12 cells was 91.2±2.6% and the printed cells aligned closely with each other forming confluent myotubes on almost all the cantilevers. Conjugated myotube and cantilever constructs responded synchronously to the electric pulses of 2 V with 40 ms duration up to 5 Hz (Fig. 3). This showed the bioprinted microdevices possessed equal

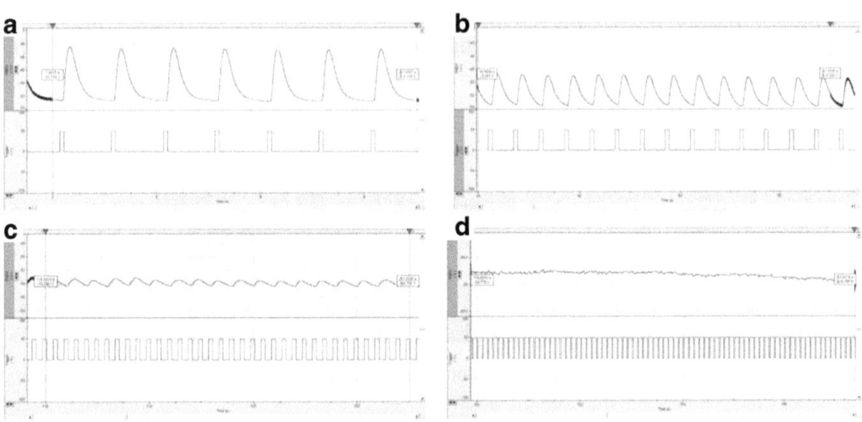

Fig. 3 Printed myotube construct responds synchronously to the applied electronic field with 2 V and 40 ms duration for each pulse. **a** 1 Hz. **b** 2 Hz. **c** 5 Hz. **d** 10 Hz.

or even better physiological properties comparing to the conventionally fabricated constructs in term of the spontaneous responses to the stimulation with significantly less culture time. Moreover, the bioprinted myotubes can also be used for muscle exercise studies with electric stimulations at various frequencies, which demonstrates the versatility of this work.

5 Cartilage Printing

Cartilage defects resulting from osteoarthritis, aging, and joint injury are a major cause of joint pain and chronic disability [53]. Mature cartilage cannot heal spontaneously because of its avascular, aneural, and alymphatic nature. The most common clinical treatments for cartilage repair include microfracture, osteochondral transfer, and autologous chondrocyte implantation. All these invasive and complicated treatments are still not able to restore the long lasting healthy cartilage [54]. Although articular cartilage was predicted to be one of the first tissues to be successfully engineered [55], the current cartilage tissue engineering strategies still cannot fabricate new tissue that is indistinguishable from native cartilage with respect to the zonal organization, extracellular matrix (ECM) composition, and mechanical properties [56]. In addition, most current cartilage repair strategies involve removing healthy cartilage tissue around the lesion site to create artificial defects for further treatment [57]. This procedure in fact causes additional necrosis to the existing healthy cartilage and leads to ultimate cartilage degeneration and failure of implanted tissue [58].

Inkjet bioprinting is able to directly repair cartilage tissue with closely mimicked native cartilage anatomy to the lesion site without additional damage. The ideal implanted tissue is expected to integrate with existing native cartilage and to repair lesions of different sizes and thicknesses. The multifaceted nature of this challenge requires a technique adaptable to variable physical dimensions and properties for tissue repair; bioprinting technology, based on inkjet printing, provides the necessary capabilities.

A standard thermal inkjet printer was modified to precisely deposit human articular chondrocytes and poly(ethylene) glycol dimethacrylate (PEGDMA; MW, 3400) layer-by-layer into a cartilage defect within an osteochondral plug for cartilage repair (Fig. 4). For a representative defect of 4 mm diameter and cartilage thickness of 2 mm, a nominal 0.23 μL of bioink estimated to contain 1140 human chondrocytes (5×10^6 cells/mL) was printed and photopolymerized for each layer to repair the cartilage defect in a layer-by-layer assembly. The thickness of each printed layer was about 18 μm. Total firing time of printhead was 1.1 s and the whole printing process completed less than 2 min. Compared to manual zonal cartilage fabrication which requires at least 11 min for UV exposure [59], bioprinting reduced UV exposure to the cells by 80 %. The viability of human chondrocytes printed with simultaneous photopolymerization increased 40 % than that when exposed to the same UV light source continuously for 10 min in manual fabrication [60].

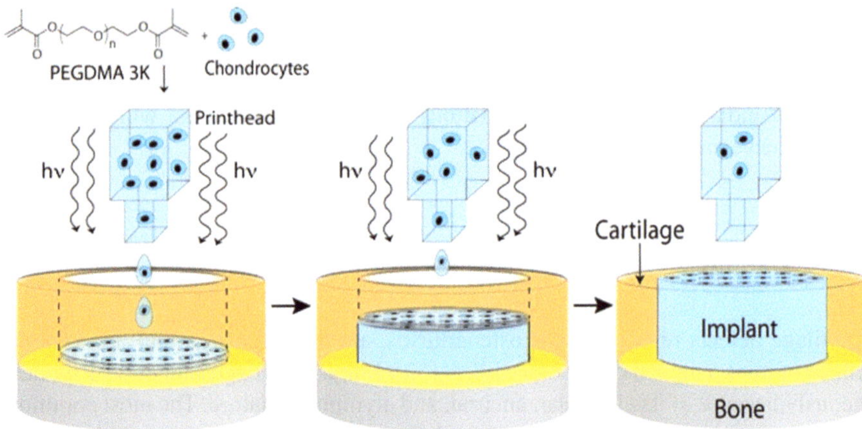

Fig. 4 Schematic of bioprinting cartilage with simultaneous photopolymerization

Printed cartilage implant attached firmly with existing tissue and greater proteoglycan deposition was also observed at the interface of implant and native cartilage. Printed cartilage in 3D biopaper had elevated glycosaminoglycan (GAG) content comparing to that without biopaper. This study indicates the importance and feasibility of direct cartilage repair and bioprinting successfully controlled placement of individual cells, preserved cell viability, maintained chondrogenic phenotype, and demonstrated integration with host tissue.

6 Bone Printing

Although bone is well known for its self-healing capacities [61], the body cannot completely heal the bone defect without intervention when it is beyond the critical size [62, 63]. Large-scale bone loss resulting from tumor resections and high impact trauma is the major cause for bone repair and implantation in clinic. The availability and functionality of bone autografts and allografts are limited to restore the normal operations. The inert implants fail over time due to repetitive loading. Therefore, tissue engineered bone which can ideally be remodeled into new bone to restore, maintain or improve its functions is becoming increasingly attractive [64].

Thermal inkjet bioprinting has been developed as an enabling technology to simultaneously deposit cells, growth factors, and biomaterial scaffolds to the desired 2D and 3D locations [10–14, 17–21]. The ejected ink drops through the nozzles are smaller than 0.03 mm in diameter, which guarantees excellent printing resolution [34, 35]. Many inkjet printed scaffolds were natural hydrogels for the enhanced biocompatibility to the cells [12, 13, 65–67]. These scaffolds usually lacked mechanical strength due to the properties of material and crosslinking methods, limiting their applications to soft tissues. Previous work also showed bone grafts created using natural hydrogels such as fibrin or alginate [68–71]. Although the cells proliferated and differentiated well in these natural hydrogels, the compressive modulus of these scaffolds is less than 5 kPa even after 4 weeks in culture, which is not ideal for bone tissue engineering [69–71].

A 3D bioprinting platform with simultaneous photopolymerization using a synthetic polymeric hydrogel was recently developed. The compressive modulus of the printed PEGDMA using layer-by-layer assembly exceeds 500 kPa, which is 100 times more than the compressive modulus of the natural hydrogels [14, 21] and in the same order of magnitude as human musculoskeletal tissue [72]. In addition, PEG hydrogel has been demonstrated to maintain cell viability and promote ECM production [14, 21, 73, 74].

Bone marrow derived stem cells are capable to migrate to the skeletal sites, proliferate and differentiate at the local injured area. Isolated human mesenchymal stem cells (hMSCs) can maintain their osteogenic potential during monolayer cell expansion *in vitro* [75]. These cells are therefore commonly used to reconstruct skeletal tissues in orthopedic tissue engineering [76–78]. hMSCs isolated from bone marrow or adipose tissue can be induced for osteogenic differentiation and form bone tissue when stimulated by ceramic scaffold [79–81]. Bioactive glass (BG) and hydroxyapatite (HA) were also reported to promote bone tissue formation [70, 82].

In bone printing, the approaches mentioned above were integrated into a novel bioprinting setup, in which hMSCs and PEGDMA combined with BG or HA or both BG and HA nanoparticles were simultaneously printed to form the homogeneous bone constructs in a layer-by-layer approach. Biochemical analysis showed significantly higher total collagen production and alkaline phosphatase (ALP) activity in hMSCs printed within PEG-HA scaffold. The higher collagen production in PEG-HA scaffold was also observed in histology studies (Fig. 5), which was consistent with the previous work by Patel et al. that HA presence increased cell ALP activity and promoted osteogenesis [83]. Collectively, HA in PEG hydrogel maintained hMSCs viability, promoted hMSCs osteogenic differentiation and biosynthetic function.

This work demonstrates the feasibility of fabricating a neobone tissue by delivering hMSCs and osteogenic factors such as HA and BG nanoparticles in strong PEG scaffold for bone tissue engineering. Using layer-by-layer assembly, the deposited hMSCs were fixed at their initially deposited positions using simultaneous photopolymerization with reduced phototoxicity. HA in scaffold significantly stimulated hMSCs osteogenic differentiation as well as osteogenic ECM production with minimal cell toxicity. Combining with previous success in cartilage bioprinting [14], it is promising to construct osteochondral interface, which is one of the most important and difficult subjects in bone tissue engineering [84].

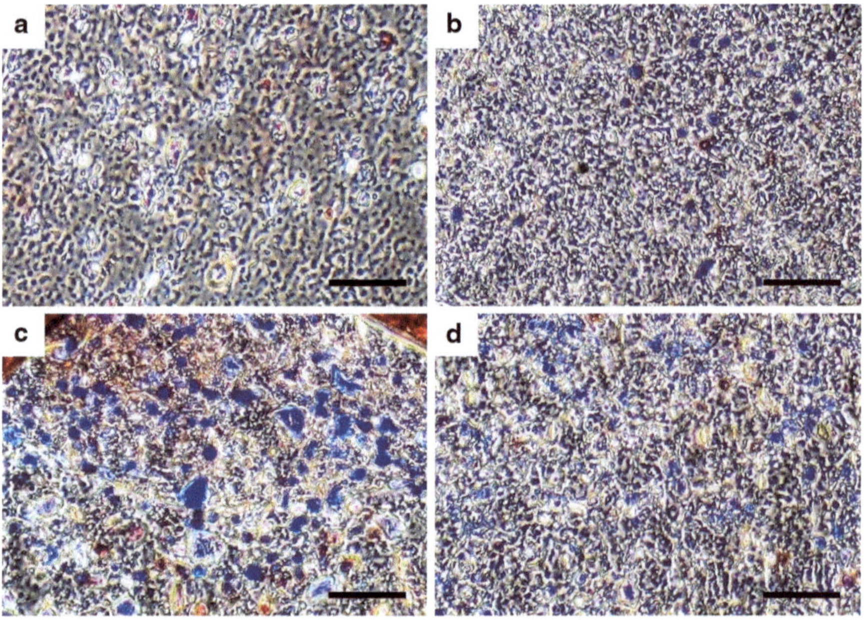

Fig. 5 Masson's trichrome staining for collagen production of hMSCs under osteogenic differentiation in various scaffolds after 21 days in culture. **a** PEG. **b** PEG-BG. **c** PEG-HA. **d** PEG-BG-HA. Scale bars: 50 μm

7 The Future

Taken together, bioprinting based on thermal inkjet printing demonstrates great feasibility of printing living systems and the flexibility of printing various subjects from soft to hard tissues with minimal side effects. In fact, the benign effects to the printed cells can be used for many other attractive applications, such as gene transfection and targeted drug delivery. The bioprinting system is versatile for 2D and 3D tissue application as well as avascular and vascular tissue construction. One promising clinical application is to develop a hand-held printer or printhead with digital control for direct tissue repair. By using 3D reconstructions of scanned lesions, bioprinting is able to precisely deliver cells, growth factors, and biomaterial scaffolds to repair the lesion with various shape and thickness. One promising direction is to combine the bioprinting approaches based on various mechanisms to meet the different challenges. Ultimately, the successful application in microvasculature fabrication also revealed the bioprinting may be the only solution to engineer thick and complex tissues with fully functional vasculature and innervation.

Acknowledgements The author would like to acknowledge Guohao Dai, Arndt F. Schilling, M.G. Finn, Kurt Breitenkamp for constructive suggestions and technical support. This work was funded by the Fundamental Research Funds for the Central Universities (WUT: 2015IB004), NSF 1011796, New York Capital Region Research Alliance grant, and Stemorgan Therapeutics R&D support (TERM002). The authors indicate no potential conflict of interest.

References

1. Langer R, Vacanti JP. Tissue engineering. Science. 1993;260(5110):920–6.
2. Auger FA, Gibot L, Lacroix D. The pivotal role of vascularization in tissue engineering. Annu Rev Biomed Eng. 2013;15:177–200.
3. Jain RK, Au P, Tam J, Duda DG, Fukumura D. Engineering vascularized tissue. Nat Biotechnol. 2005;23(7):821–3.
4. Atala A, Bauer SB, Soker S, Yoo JJ, Retik AB. Tissue-engineered autologous bladders for patients needing cystoplasty. Lancet. 2006;367(9518):1241–6.
5. Levenberg S, Rouwkema J, Macdonald M, Garfein ES, Kohane DS, Darland DC, et al. Engineering vascularized skeletal muscle tissue. Nat Biotechnol. 2005;23(7):879–84.
6. Ma PX, Choi JW. Biodegradable polymer scaffolds with well-defined interconnected spherical pore network. Tissue Eng. 2001;7(1):23–33.
7. Kang HW, Park JH, Kang TY, Seol YJ, Cho DW. Unit cell-based computer-aided manufacturing system for tissue engineering. Biofabrication. 2012;4(1):015005.
8. Hu C, Uchida T, Tercero C, Ikeda S, Ooe K, Fukuda T, et al. Development of biodegradable scaffolds based on magnetically guided assembly of magnetic sugar particles. J Biotechnol. 2012;159(1–2):90–8. (Feb 14).
9. Nerem RM, Seliktar D. Vascular tissue engineering. Annu Rev Biomed Eng. 2001;3:225–43.
10. Cui X, Boland T, D'Lima DD, Lotz MK. Thermal inkjet printing in tissue engineering and regenerative medicine. Recent Pat Drug Deliv Formul. 2012;6(2):149–55.
11. Cui X, Dean D, Ruggeri ZM, Boland T. Cell damage evaluation of thermal inkjet printed Chinese hamster ovary cells. Biotechnol Bioeng. 2010;106(6):963–9.
12. Cui X, Boland T. Human microvasculature fabrication using thermal inkjet printing technology. Biomaterials. 2009;30(31):6221–7.
13. Cui X, Boland T. Simultaneous deposition of human microvascular endothelial cells and biomaterials for human microvasculature fabrication using inkjet printing. NIP24/digital Fabrication 2008: 24th International Conference on Digital Printing Technologies, Technical Program and Proceedings 2008;24:480–3.
14. Cui X, Breitenkamp K, Finn MG, Lotz M, D'Lima DD. Direct human cartilage repair using three-dimensional bioprinting technology. Tissue Eng Part A. 2012;18(11–12):1304–12.
15. Gao G, Yonezawa T, Hubbell K, Dai G, Cui X. Inkjet-bioprinted acrylated peptides and PEG hydrogel with human mesenchymal stem cells promote robust bone and cartilage formation with minimal printhead clogging. Biotechnol J. 2015. Jan 8 doi:10.1002/biot.201400635.
16. Gao G, Schilling AF, Yonezawa T, Wang J, Dai G, Cui X. Bioactive nanoparticles stimulate bone tissue formation in bioprinted three-dimensional scaffold and human mesenchymal stem cells. Biotechnol J. 2014;9(10):1304–11.
17. Cui X, Breitenkamp K, Finn MG, Lotz M, Colwell CW Jr. Direct human cartilage repair using thermal inkjet printing technology. Osteoarthritis Cartilage. 2011;19:S47–S8.
18. Cui X, Gao G, Yonezawa T, Dai G. Human cartilage tissue fabrication using three-dimensional inkjet printing technology. J Vis Exp 2014;(88), e51294. doi:10.3791/51294.
19. Cui X, Gao G, Qiu Y. Accelerated myotube formation using bioprinting technology for biosensor applications. Biotechnol Lett. 2013;35(3):315–21.
20. Cui X, Hasegawa A, Lotz M, D'Lima D. Structured three-dimensional co-culture of mesenchymal stem cells with meniscus cells promotes meniscal phenotype without hypertrophy. Biotechnol Bioeng. 2012;109(9):2369–80.
21. Cui X, Breitenkamp K, Lotz M, D'Lima D. Synergistic action of fibroblast growth factor-2 and transforming growth factor-beta1 enhances bioprinted human neocartilage formation. Biotechnol Bioeng. 2012;109(9):2357–68.
22. Cohen DL, Malone E, Lipson H, Bonassar LJ. Direct freeform fabrication of seeded hydrogels in arbitrary geometries. Tissue Eng. 2006;12(5):1325–35.
23. Iwami K, Noda T, Ishida K, Morishima K, Nakamura M, Umeda N. Bio rapid prototyping by extruding/aspirating/refilling thermoreversible hydrogel. Biofabrication. 2010;2(1):014108.

24. Shor L, Guceri S, Chang R, Gordon J, Kang Q, Hartsock L, et al. Precision extruding deposition (PED) fabrication of polycaprolactone (PCL) scaffolds for bone tissue engineering. Biofabrication. 2009;1(1):015003.
25. Barron JA, Wu P, Ladouceur HD, Ringeisen BR. Biological laser printing: A novel technique for creating heterogeneous 3-dimensional cell patterns. Biomed Microdevices. 2004;6(2):139–47.
26. Guillemot F, Souquet A, Catros S, Guillotin B, Lopez J, Faucon M, et al. High-throughput laser printing of cells and biomaterials for tissue engineering. Acta Biomater. 2010;6(7):2494–500.
27. Guillemot F, Souquet A, Catros S, Guillotin B. Laser-assisted cell printing: principle, physical parameters versus cell fate and perspectives in tissue engineering. Nanomedicine (Lond). 2010;5(3):507–15.
28. Mironov V, Visconti RP, Kasyanov V, Forgacs G, Drake CJ, Markwald RR. Organ printing: tissue spheroids as building blocks. Biomaterials. 2009;30(12):2164–74.
29. Moon S, Hasan SK, Song YS, Xu F, Keles HO, Manzur F, et al. Layer by layer three-dimensional tissue epitaxy by cell-laden hydrogel droplets. Tissue Eng Part C Methods 2010;16(1):157–66.
30. Odde DJ, Renn MJ. Laser-guided direct writing for applications in biotechnology. Trends Biotechnol. 1999;17(10):385–9.
31. Odde DJ, Renn MJ. Laser-guided direct writing of living cells. Biotechnol Bioeng. 2000;67(3):312–8.
32. Mohebi MM, Evans JRG. A drop-on-demand ink-jet printer for combinatorial libraries and functionally graded ceramics. J Comb Chem. 2002;4(4):267–74.
33. Beeson R. Thermal (TIJ) or Piezo? Who cares? IMI 7th Annual Ink Jet Printing Conference; 1998.
34. Hock SW, Johnson DA, Van Veen MA. Inventors; Print quality optimization for a color ink-jet printer by using a larger nozzle for the black ink only.US5521622. 1996.
35. Canfield B, Clayton H, Yeung KWW. Inventors; Method and apparatus for reducing the size of drops ejected from a thermal ink jet printhead.US5673069. 1997.
36. Hudson KR, Cowan PB, Gondek JS. Inventors; Ink drop volume variance compensation for inkjet printing.US6042211. 2000.
37. de Jong J, de Bruin G, Reinten H, van den Berg M, Wijshoff H, Versluis M, et al. Air entrapment in piezo-driven inkjet printheads. J Acoust Soc Am. 2006 ;120(3):1257–65.
38. Sirringhaus H, Kawase T, Friend RH, Shimoda T, Inbasekaran M, Wu W, et al. High-resolution inkjet printing of all-polymer transistor circuits. Science. 2000;290(5499):2123–6.
39. Okamoto T, Suzuki T, Yamamoto N. Microarray fabrication with covalent attachment of DNA using Bubble Jet technology. Nat Biotechnol. 2000;18(4):438–41.
40. Goldmann T, Gonzalez JS. DNA-printing: utilization of a standard inkjet printer for the transfer of nucleic acids to solid supports. J Biochem Biophys Methods. 2000;42(3):105–10.
41. Seetharam R, Sharma SK. Purification and analysis of recombinant proteins. New York: Marcel Dekker; 1991. p. 69.
42. Tirella A, Vozzi F, De MC, Vozzi G, Sandri T, Sassano D, et al. Substrate stiffness influences high resolution printing of living cells with an ink-jet system. J Biosci Bioeng. 2011;112(1):79–85.
43. Xu T, Rohozinski J, Zhao W, Moorefield EC, Atala A, Yoo JJ. Inkjet-mediated gene transfection into living cells combined with targeted delivery. Tissue Eng Part A. 2009;15(1):95–101.
44. Catros S, Guillemot F, Nandakumar A, Ziane S, Moroni L, Habibovic P, et al. Layer-by-layer tissue microfabrication supports cell proliferation in vitro and in vivo. Tissue Eng Part C Methods. 2012;18(1):62–70.
45. Xi J, Schmidt JJ, Montemagno CD. Self-assembled microdevices driven by muscle. Nat Mater. 2005;4(2):180–4.
46. Tanaka Y, Sato K, Shimizu T, Yamato M, Okano T, Kitamori T. A micro-spherical heart pump powered by cultured cardiomyocytes. Lab Chip. 2007;7(2):207–12.

47. Harms H, Wells MC, van dM Jr. Whole-cell living biosensors–are they ready for environmental application? Appl Microbiol Biotechnol. 2006;70(3):273–80.
48. Bers DM. Cardiac excitation-contraction coupling. Nature. 2002;415(6868):198–205.
49. Asano T, Ishizua T, Yawo H. Optically controlled contraction of photosensitive skeletal muscle cells. Biotechnol Bioeng. 2012;109(1):199–204.
50. Yaffe D, Saxel O. A myogenic cell line with altered serum requirements for differentiation. Differentiation. 1977;7(3):159–66.
51. Miller JB. Myogenic programs of mouse muscle cell lines: expression of myosin heavy chain isoforms, MyoD1, and myogenin. J Cell Biol. 1990;111(3):1149–59.
52. Fujita H, Shimizu K, Nagamori E. Novel method for measuring active tension generation by C2C12 myotube using UV-crosslinked collagen film. Biotechnol Bioeng. 2010;106(3):482–9.
53. Mow VC, Hayes WC. Basic orthopaedic biomechanics. 2 ed. Philadelphia: Lippincott Williams & Wilkins; 1997.
54. Rasanen P, Paavolainen P, Sintonen H, Koivisto AM, Marja B, Ryynanen OP, et al. Effectiveness of hip or knee replacement surgery in terms of quality-adjusted life years and costs. Acta Orthop. 2007;78(1):108–15.
55. Brittberg M, Lindahl A, Nilsson A, Ohlsson C, Isaksson O, Peterson L. Treatment of deep cartilage defects in the knee with autologous chondrocyte transplantation. N Engl J Med. 1994;331(14):889–95.
56. Hunziker EB. Articular cartilage repair: basic science and clinical progress. A review of the current status and prospects. Osteoarthritis Cartilage. 2002;10(6):432–63.
57. Kalson NS, Gikas PD, Briggs TWR. Current strategies for knee cartilage repair. Int J Clin Pract. 2010;64(10):1444–52.
58. Shapiro F, Koide S, Glimcher MJ. Cell origin and differentiation in the repair of full-thickness defects of articular-cartilage. J Bone Joint Surg-Am. 1993;75 A(4):532–53.
59. Kim TK, Sharma B, Williams CG, Ruffner MA, Malik A, McFarland EG, et al. Experimental model for cartilage tissue engineering to regenerate the zonal organization of articular cartilage. Osteoarthritis Cartilage. 2003;11(9):653–64.
60. Cui X, Breitenkamp K, Finn MG, Lotz MK, D'Lima DD. Direct human cartilage repair using 3D bioprinting technology. Tissue Eng Part A. 2012;18(11–12):1304–12.
61. Mourino V, Boccaccini AR. Bone tissue engineering therapeutics: controlled drug delivery in three-dimensional scaffolds. J R Soc Interface. 2010;7(43):209–27.
62. Jones AC, Arns CH, Sheppard AP, Hutmacher DW, Milthorpe BK, Knackstedt MA. Assessment of bone ingrowth into porous biomaterials using MICRO-CT. Biomaterials. 2007;28(15):2491–504.
63. Seitz H, Rieder W, Irsen S, Leukers B, Tille C. Three-dimensional printing of porous ceramic scaffolds for bone tissue engineering. J Biomed Mater Res B Appl Biomater. 2005;74(2):782–8.
64. Rezwan K, Chen QZ, Blaker JJ, Boccaccini AR. Biodegradable and bioactive porous polymer/inorganic composite scaffolds for bone tissue engineering. Biomaterials. 2006;27(18):3413–31.
65. Deitch S, Kunkle C, Cui X, Boland T, Dean D. Collagen matrix alignment using inkjet printer technology. Mater Res Soc Symp Proc. 2008;1094:52–7.
66. Boland T, Xu T, Damon B, Cui X. Application of inkjet printing to tissue engineering. Biotechnol J. 2006;1(9):910–7.
67. Boland T, Cui X, Aho M, Baicu C, Zile M. Image based printing of structured biomaterials for realizing complex 3D cardiovascular constructs. J Imaging Sci Technol. 2006;2:86–8.
68. Catelas I, Sese N, Wu BM, Dunn JC, Helgerson S, Tawil B. Human mesenchymal stem cell proliferation and osteogenic differentiation in fibrin gels in vitro. Tissue Eng. 2006;12(8):2385–96.
69. Spalazzi JP, Dagher E, Doty SB, Guo XE, Rodeo SA, Lu HH. In vivo evaluation of a multiphased scaffold designed for orthopaedic interface tissue engineering and soft tissue-to-bone integration. J Biomed Mater Res A. 2008;86(1):1–12.

70. Khanarian NT, Jiang J, Wan LQ, Mow VC, Lu HH. A hydrogel-mineral composite scaffold for osteochondral interface tissue engineering. Tissue Eng Part A. 2011;18(5–6):533–45. Nov 8.
71. Spalazzi JP, Doty SB, Moffat KL, Levine WN, Lu HH. Development of controlled matrix heterogeneity on a triphasic scaffold for orthopedic interface tissue engineering. Tissue Eng. 2006;12(12):3497–508.
72. Hoenig E, Winkler T, Mielke G, Paetzold H, Schuettler D, Goepfert C, et al. High amplitude direct compressive strain enhances mechanical properties of scaffold-free tissue-engineered cartilage. Tissue Eng Part A. 2011;17(9–10):1401–11.
73. Bryant SJ, Anseth KS. Hydrogel properties influence ECM production by chondrocytes photoencapsulated in poly(ethylene glycol) hydrogels. J Biomed Mater Res. 2002;59(1):63–72.
74. Elisseeff J, McIntosh W, Anseth K, Riley S, Ragan P, Langer R. Photoencapsulation of chondrocytes in poly(ethylene oxide)-based semi-interpenetrating networks. J Biomed Mater Res. 2000;51(2):164–71.
75. Bruder SP, Jaiswal N, Haynesworth SE. Growth kinetics, self-renewal, and the osteogenic potential of purified human mesenchymal stem cells during extensive subcultivation and following cryopreservation. J Cell Biochem. 1997;64(2):278–94.
76. Caplan AI, Bruder SP. Mesenchymal stem cells: building blocks for molecular medicine in the 21st century. Trends Mol Med. 2001;7(6):259–64.
77. Triffitt JT. Osteogenic stem cells and orthopedic engineering: summary and update. J Biomed Mater Res. 2002;63(4):384–9.
78. Oreffo RO, Triffitt JT. Future potentials for using osteogenic stem cells and biomaterials in orthopedics. Bone 1999;25(2 Suppl):5S–9 S.
79. Leboy PS, Beresford JN, Devlin C, Owen ME. Dexamethasone induction of osteoblast mRNAs in rat marrow stromal cell cultures. J Cell Physiol. 1991;146(3):370–8.
80. Rickard DJ, Kassem M, Hefferan TE, Sarkar G, Spelsberg TC, Riggs BL. Isolation and characterization of osteoblast precursor cells from human bone marrow. J Bone Miner Res. 1996;11(3):312–24.
81. Kon E, Muraglia A, Corsi A, Bianco P, Marcacci M, Martin I, et al. Autologous bone marrow stromal cells loaded onto porous hydroxyapatite ceramic accelerate bone repair in critical-size defects of sheep long bones. J Biomed Mater Res. 2000;49(3):328–37.
82. Jiang J, Tang A, Ateshian GA, Guo XE, Hung CT, Lu HH. Bioactive stratified polymer ceramic-hydrogel scaffold for integrative osteochondral repair. Ann Biomed Eng. 2010;38(6):2183–96.
83. Patel M, Patel KJ, Caccamese JF, Coletti DP, Sauk JJ, Fisher JP. Characterization of cyclic acetal hydroxyapatite nanocomposites for craniofacial tissue engineering. J Biomed Mater Res A. 2010;94(2):408–18.
84. Hunziker EB, Driesang IM. Functional barrier principle for growth-factor-based articular cartilage repair. Osteoarthritis Cartilage. 2003;11(5):320–7.

Bioprinting of Dynamic Human Organs-on-Chips: Enabling Technologies for Rapid Drug Development and Personalized Medicine

Dileep Daniel Monie and Sujata Kumari Bhatia

1 Clinical Need for Organs-on-Chips

1.1 History of Drug Discovery and Development

Medicine aims to alleviate the human suffering caused by disease. The approaches employed in this process have evolved over time. Ancient healers used herbal and folk remedies—established through uncontrolled experiments and unblinded observations of drug effects on the ill. Knowledge of these treatments were passed down through generations and shared between communities, likely facilitated by trade across borders. These therapies did not cost much to develop, but they came about slowly over millennia, put human lives at relatively high risk, and often worked no better than a placebo.

In stark contrast, modern medicines are a product of the scientific method. Hypotheses are thoughtfully generated in accordance with meticulous observations. These are then rigorously tested, data recorded in great detail, and the results are subjected to objective statistical analyses. This approach has served humanity well for a number of years. Active pharmaceutical ingredients (APIs) for a variety of indications have been identified and translated to the clinic. The vast majority of these owe their success to prior art, having been derived from the phytochemicals of traditional medicines, or to sheer serendipity.

The genomic era promised a new age of personalized, molecular medicines. Indeed, a few rationally designed compounds such as imatinib, which targets the

S. K. Bhatia (✉)
School of Engineering and Applied Sciences, Harvard University, Cambridge, MA, USA
e-mail: sbhatia@seas.harvard.edu

D. D. Monie
Faculty of Arts and Sciences, Harvard University, Cambridge, MA, USA
e-mail: dmonie@post.harvard.edu

K. Turksen (ed.), *Bioprinting in Regenerative Medicine,* Stem Cell Biology and Regenerative Medicine, DOI 10.1007/978-3-319-21386-6_6

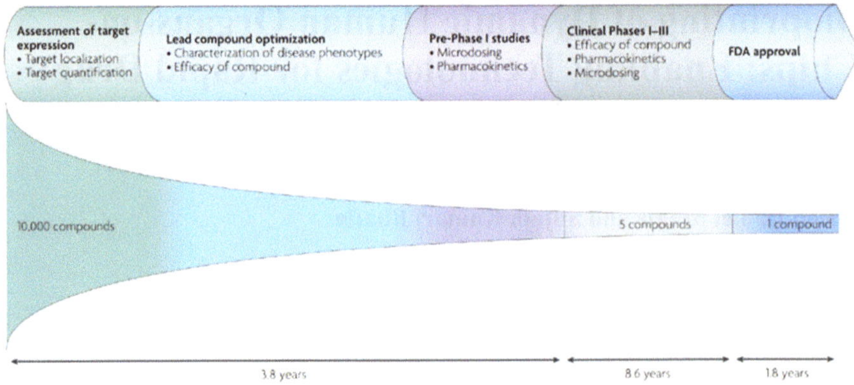

Fig. 1 This is a typical timeline for screening a compound library to produce a single marketed drug. Note that the time axis is not to scale. Figure is adapted from [3]

bcr-abl fusion tyrosine kinase in chronic myelogenous leukemia (CML) [1], have lent credence to the concept. But the overall trends in the pharmaceutical industry suggest that these few compounds may be the exception rather than the rule. A research report by the InnoThink Center For Research In Biomedical Innovation and Thomson Reuters Fundamentals via FactSet Research Systems shows that the average drug approved by the United States Food and Drug Administration (FDA) between 1997 and 2011 cost over $ 4 billion [2]. Willmann et al. reviewed an earlier but similar data set—their results suggest that the mean time to bring a drug from the bench to the bedside is 14.2 years [3]. A closer look at the drug discovery and development process highlights the bottlenecks and inefficiencies of the current paradigm (Fig. 1).

For small molecules, the approach starts with a large compound library that is curated to maximize chemical diversity and adhere to Lipinski's rule-of-five (RO5) for druglikeness, which "predicts that poor absorption or permeation is more likely when there are more than 5H-bond donors, 10H-bond acceptors, the molecular weight is greater than 500 and the calculated Log P is greater than 5" [4]. These compounds or biomolecules are then subjected to high-throughput screening (HTS) assays that have been designed for a particular disease phenotype or disease-associated molecular target.

Larger biomolecules, such as peptides and monoclonal antibodies (mAbs), are initially isolated from biological materials then modified to generate some diversity using genetic and protein engineering methods. These are also screened using laboratory assays but often on a smaller scale. This is likely due to the cost and complexity of creating new biomolecules combined with the high level of target specificity and control over design inherent in the technology.

Modern assays are trending toward molecular targets—likely because of the lower upfront cost and perceived increase in specificity—that have been identified and validated by the latest research. Molecular target assays can be either biochemical or cell-based, the latter of which is often lower throughput but more informative. Phenotypic assays, however, are still prevalent, especially as a secondary screening tool. After all, it is the phenotype that must be altered in order to ameliorate the disease state.

After the primary and secondary screens, hits are rank ordered by their effectiveness or inhibition concentrations (EC_{50} or IC_{50}). These hits are often counter-screened to establish cytotoxicity concentrations (CC_{50}), revealing false-positives due assay artifacts and hinting at potential clinical toxicities. These data collectively inform the development of a structure-activity relationship (SAR) model, which is used by medicinal chemists to iteratively modify compounds in a rational manner. Standardized *in vitro* bioassays to predict absorption, distribution, metabolism, excretion, and toxicity (ADMET) are usually initiated at this point.

Lead-like chemical structures are nominated from these data and are further optimized through medicinal chemistry efforts. The resulting lead compounds—with novel, patentable compositions of matter—that meet specified activity and predicted ADMET requirements are pushed into preclinical animal studies. The ability of the lead compounds to modulate animal models of disease, along with their *in vivo* pharmacokinetics (PK) and pharmacodynamics (PD) properties, allow for further whittling down of candidates. Currently, the FDA requires preclinical toxicity data from testing in two different animal models, including at least one nonrodent [5].

The leads with the best performance profiles in animal studies are nominated for clinical development and are awarded investigational new drug (IND) status by the FDA. Analysis by Hay et al. suggests that only about 15 % of these INDs will have their new drug applications (NDAs) or biologic license applications (BLAs) approved for commercialization [6]. The same study shows that over half of the failures are due to insufficient efficacy and a third are due to safety concerns.

Given that the majority of the costs for developing new drugs are incurred in the clinical stages—and that the large number of clinical failures account for the stratospheric cost per new drug—more predictive preclinical assays are needed. Identifying poorly efficacious or potentially toxic compounds earlier will speed up the drug development process while also minimizing costs due to clinical stage failures (Fig. 2) [7]. Perhaps most importantly, however, is that better assays will lessen the risks posed to trial volunteers and patients.

1.2 Tissue Culture Bioassays

The bioassays that inform critical decisions in today's pharmaceutical industry are rooted in tissue culture. Ever since the first human cells were cultured *in vitro*—a cervical cancer adenocarcinoma from Henrietta Lacks in 1951 [8]—researchers have been able to test compounds on human cells before introducing them in to living persons. This ushered in an era of optimism: *in vitro* human tissues could provide insight into the efficacy and toxicities of novel compounds. Furthermore, tissue culture is relatively inexpensive and more amenable to HTS than using ani-

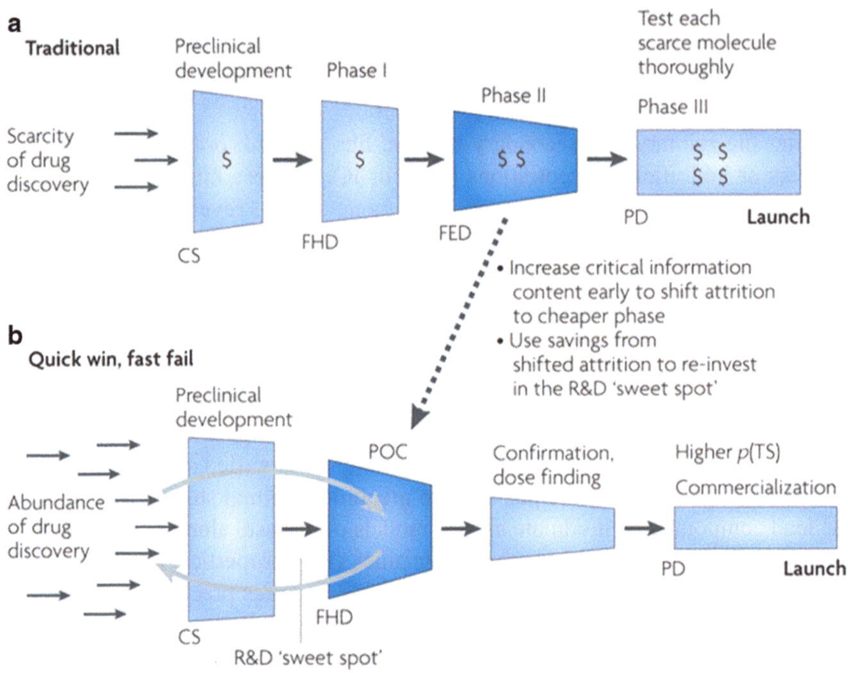

Fig. 2 **a** The width or the blue regions represent the costs associated with each phase of the current paradigm of drug development. **b** Failing early increases the efforts and cost of preclinical studies but lowers the overall cost of the process while increasing the rate of new drug approvals. Figure is adapted from [7]

mals; it also serves as the basis for high-content screening (HCS). As a result, more compounds can be screened faster and cheaper, all while generating more data that can inform decision making.

Two broad types of cell-based screens are used for evaluating the suitability of a drug for a particular disease indication. Traditionally, phenotypic assays simply looked at whether or not a compound could alter the phenotype of a diseased cell in such a manner as to restore normal function. While phenotype restoration in the patient is the ultimate goal of a successful therapeutic, the cellular phenotype is not necessarily representative of the body phenotype. This type of assay may also inadvertently miss or even promote toxic off-target effects, while failing to take advantage of research on the molecular basis of disease.

The other major type of cellular assay is the molecular assay—usually a cell line genetically engineered with a reporter construct. For example, if the drug target is a transcription factor, a downstream promoter may be linked to a green fluorescent protein (GFP) gene. An effective compound could be defined by its ability to modulate GFP expression. Such an assay allows for screening of a compound's effects on a single biomolecule or signaling pathway, often yielding highly specific interactions. But this type of assay also abstracts more complex biological systems that play a role in disease and is further removed from the disease phenotype.

Beyond these activity assays, a number of standardized assays have been developed to predict ADMET. A classic example is the Caco-2 transwell assay, which is used to predict intestinal absorption of orally delivered drugs [9]. This is a tissue culture system that attempts to approximate the human gut. It consists of two liquid chambers—apical and basolateral—separated by a permeable membrane. Caco-2 human colorectal adenocarcinoma cells are cultured on the membrane where they form a monolayer and polarize. A known concentration of compound is added to the apical chamber, which represents the small intestine lumen. The system is incubated for several hours until equilibrium is reached. Compound concentrations are then determined both in the apical chamber and in the basolateral chamber, which represents the interstitial fluid between the basement membrane and the capillaries. The vectorial transport ratio is calculated using these concentrations:

$$vectorial\ transport \frac{[apical]}{[basolateral]}$$

If this ratio is significantly greater than 1, the intestinal absorption is predicted to be low. If it is significantly less than 1, the intestinal absorption is predicted to be high.

But the optimism generated by tissue culture assays has proved premature—these tissue cultures are poor at predicting drug effects on normal and diseased human physiologies. Even the relatively complicated Caco-2 system is unable to actively transport compounds like an actual small intestine [10, 11]. Three-dimensional (3D) culture systems have been developed to better mimic the human condition, but 3D cultures to date have not fared much better than their two-dimensional (2D) counterparts [12]. This is likely due to a number of reasons, the most salient of which are: (1) cancerous or immortalized cells are fundamentally different than normal cells; (2) monolayers or monocultures of cells do not have the same shape and interactions with their milieu as *in situ*; (3) dynamic motions and mechanical forces experienced *in vivo* are not recapitulated *in vitro*; and (4) tissue cultures in isolation are not subjected to endocrine hormones and other signals from distant cells.

1.3 Animal Models

Animal models address these issues: (1) they are comprised of normal cells; (2) the cells are assembled into 3D tissues and organs; (3) these tissues and organs experience physiologic dynamic motions and mechanical forces; and (4) these tissues and organs also interact with distant tissues through lymphatic and circulatory systems. But, again, animals clearly have drawbacks as evidenced by the current state of drug development. In additional to issues of throughput, cost, study time, and the ethics of animal testing, animal models just are not representative of human biology.

The National Human Genome Research Institute notes that "on average, the protein-coding regions of the mouse and human genomes are 85 % identical; some genes are 99 % identical while others are only 60 % identical" [13]. These statistics do not include the majority of the genome, which do not encode proteins but are thought to play critical roles in gene regulation and genome stability [14]. So while it may be reasonable to believe that a 99 % homologous, isolated mouse protein interacts with a particular compound just like the human homolog does, it is highly unlikely that entire signaling pathways or other biological systems will behave similarly. At the genotype and phenotype levels, there is simply no substitution for the human body.

1.4 Advent of Organs-on-Chips

The design criteria for engineering the missing preclinical link are clear: (1) a solution must be comprised of normal human cells; (2) cells of different types must be co-cultured in a 3D geometry that recapitulates the *in vivo* milieu; (3) these cultures must experience physiological dynamic motions, fluid flows, spatiotemporal gradients, and mechanical forces; and (4) engineered tissues and organs should be connected together in a manner that simulates human biology. Organs-on-chips are microphysiological systems (MPS) that aim to meet these goals. By leveraging microfabrication techniques to position cells in three dimensions, exert mechanical forces, and circulate fluids, scale models of human organ systems can provide critical insight into pharmaceutical actions.

The development of organs-on-chips from conventional cell culture was a gradual process. Huh, Hamilton, and Ingber detail this evolution in [15]. They point out that while gel-based 3D cell cultures offered more representative morphologies, these "fail to reconstitute features of living organs that are crucial for their function, including tissue–tissue interfaces, spatiotemporal gradients of chemicals and oxygen, and the mechanically active microenvironment" integral to the living body. Observations that 2D cultures of rat liver [16], capillary [17], and smooth muscle [18] cells behave more like their *in vivo* counterparts when restricted in growth provided the impetus for applying microfluidic technologies to cell cultures.

One of the first successful designs of an organ-on-a-chip is an artificial liver sinusoid, the construction of which was published in 2007 [19]. This device introduces physiologic fluid flows, cell-cell interactions, and dynamic nutrient gradients to a 3D culture of hepatocytes. As a result, it is able to accurately predict the long-term hepatotoxicity of the anti-inflammatory compound diclofenac. The authors of this study also point out the feasibility for multiplexing these chips to increase throughput.

This design has been followed by several other MPS, such as bone marrow-on-a-chip [20] and kidney-on-a-chip [21]. An advance made by some new generation devices, namely a breathing lung-on-a-chip [22], peristaltic gut-on-a-chip [23], and beating heart-on-a-chip [24], is that they exert mechanical forces and permit motion to replicate dynamic human physiology. This has facilitates formation of cell morphologies consistent with that seen in human tissue histology. In the case of the gut-on-a-chip, intestinal microbiota also play a role in the development and function of the MPS.

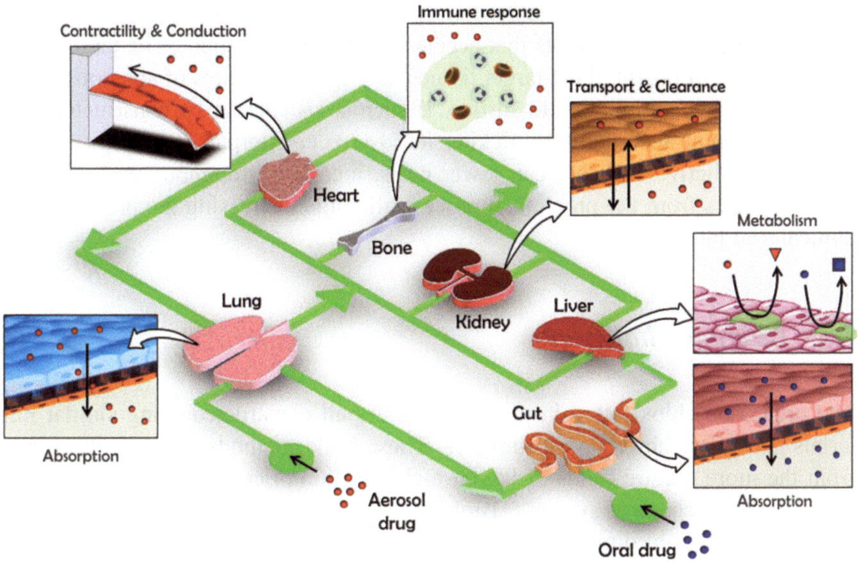

Fig. 3 A proposed design for a human-on-a-chip, which is comprised of several different organs-on-chips linked together in a microfluidic circuit. In this example, an oral drug (*blue dots*) is administered to a gut-on-a-chip, which absorbs the compound and sends it to a liver-on-a-chip for metabolism. The metabolites are then distributed to the other organs-on-chips, which are monitored for target and off-target effects. Alternatively, an aerosol drug (*red dots*) is administered to a lung-on-a-chip, where it is absorbed and enters systemic circulation. In such a system, the kidney-on-a-chip could be used to predict urinary excretion. To make the human-on-a-chip even more realistic, the bone marrow-on-a-chip can be used to produce the blood that is circulated through the entire system. Figure is adapted from [15]

The next step is to take these chips and link them together in a microfluidic circuit. Huh, Hamilton, and Ingber call this concept the human-on-a-chip, which is shown in Fig. 3 [15]. An oral drug could be administered to a gut-on-a-chip, which absorbs the compound and sends it to a liver-on-a-chip for metabolism. The metabolites are then distributed to the other organs-on-chips, which are monitored for target and off-target effects. Alternatively, an aerosol drug could be administered to a lung-on-a-chip, where it is absorbed and directly enters the systemic circulation. A kidney-on-a-chip could be used to predict urinary excretion. To make the human-on-a-chip even more realistic, the bone marrow-on-a-chip can be used to produce the blood that is circulated through the entire system.

It is important to recognize that each organ-on-a-chip does not aim to recapitulate the entire organ, just its functional subunits. For example, the lung-on-a-chip is actually just a replica of the parenchymal alveolar-capillary interface; it abstracts away the supporting stromal tissues. This is because the supporting structures theoretically add complexity and bulk without improving the predictive power of the device. In addition, entire organs probably cannot be miniaturized significantly without violating the square-cube law [25] and thus altering the biomechanics of the model system.

Similarly, the human-on-a-chip concept likely does not need to include every organ in order to offer insights that accelerate drug development. The target organ plus a sample of potential off-target tissues, along with the sites of absorption, metabolism, and excretion should suffice. If such a system can be constructed with each component accurately mimicking its *in vivo* counterpart, the result may be a highly predictive tool that can lower costs, increase throughput, and eliminate animal models from the preclinical process all while decreasing the probability of clinical trial failures.

1.5 Current Challenges

The organs-on-chips found in the literature are all constructed using similar materials and microfabrication techniques. These build methods are often borrowed from the more mature fields of lab-on-a-chip (LOC) microfluidics and biological microelectromechanical systems (BioMEMS), most commonly involving the soft lithography of poly(dimethylsiloxane) (PDMS) elastomer. This biomaterial has physical properties that make it useful constructing organs-on-chips. It is optically transparent, enabling visual observation, and acts as an electrical insulator [26], allowing currents to naturally pass through cells. The elastic modulus is quite tunable, which is essential for emulating the movement—either through pneumatic actuation or passive compliance—of diverse tissues ranging from nerve and muscle [27] to vasculature [28]. The surface chemistry of PDMS can be functionalized to adjust hydrophobicity with oxygen plasma treatment [29] and cell adhesion with extracellular matrix (ECM) coating [30]. Micropatterning of surface treatments allows for sophisticated fluidic and cellular control [31–33].

Despite the major advantages that the current generation of organs-on-chips offers over conventional drug development tools, several technical hurdles and bottlenecks remain that hinder their path to widespread adoption and regulatory acceptance. The largest problem with PDMS-based microfluidics is bubble formation [34], which obstructs fluid flows, damages the cellular microenvironment, and causes unpredictable behavior of the device [35]. Other major challenges include the small molecule absorption and permeability of PDMS [34], and the time and cost of creating modified designs for research and personalized medicine.

The flexibility of modern 3D bioprinting technologies facilitates solutions to some of these issues. Replica molding of PDMS using 3D printed molds speeds up the engineering design-build-test cycles by permitting rapid prototyping [36]. This in turn lowers the time needed for troubleshooting organ-on-a-chip designs and experimenting with novel biomaterials. Future bioprinting techniques may also speed up basic biomedical research by allowing scientists to replace their traditional cell cultures with on-demand MPSs fine-tuned to their projects. Finally, a bioprinting approach to fabrication makes it feasible to tailor organs-on-chips to individual patients at the point of care. This opens the door to the next generation of clinical diagnostics required for personalized medicine.

2 Bioprinting Design and Fabrication

2.1 Design

Like any fabrication technique, bioprinting methodologies offer both advantages and constraints that must be taken into consideration when designing organs-on-chips. A major advantage is that nearly any shape that can be rendered in 3D on a computer screen and sliced into layers can be printed. Some more complex designs, however, may require significant post-processing of the print. This may involve manual removal of supporting struts or assembly of multiple, discretely printed components. Another design consideration is the intended production scale of the final product. Large scale printing many not be feasible so alternative fabrication methods, such as tradition injection molding, could be required [37]. In such a case, the limitations of those methods will also play a role in design decisions.

Minor alterations in a design may result in drastic changes to the post-processing requirements. The need for additional components to be printed and assembled could limit what modifications can be made when tailoring organs-on-chips to individual patients at the point of care. An example of this may be modifying the design of the lung-on-a-chip to emulate the hemodynamics of a patient with pulmonary arterial hypertension (PAH). In addition to elevated blood pressures, there is turbulent flow resulting from plexiform lesions [38]. These intimal obstructions invade the arterial lumen, typically at branch points. Depending on the diagnostic or therapeutic goals, part or all of such a lesion must be introduced into the design, likely requiring extra bioprinted components.

Given the intent to use these organs-on-chips in preclinical development—in addition to or in lieu of animal models—as well as in the clinic, regulatory bodies may impose their own design constraints on the devices. Part 58 of the Code of Federal Regulations (CFR) Title 21 explains the Good Laboratory Practice for Nonclinical Laboratory Studies (GLP) requirements [39]. Any equipment using in the fabrication, including 3D bioprinters, must be designed, maintained, and calibrated in compliance with subpart D. The characterization and handling of the bioprinted organs-on-chips will then be subject to subpart F. The language is sufficiently vague that there are likely no insurmountable hurdles; however, the inherent variability that comes with custom bioprinting tissues needs to be addressed in a manner that facilitates drug development and personalized medicine without compromising data integrity and patient safety.

Design ultimately hinges on the fabrication methods utilized and, especially in the case of bioprinting, the choice of biomaterials. The availability of suitable, well-characterized biomaterials is perhaps the single largest challenge with the bioprinting of organs-on chips. To date, not many flexible biocompatible materials that are amenable to direct 3D printing have been identified. Those available commercially, such as the Stratasys PolyJet MED610 photopolymer [40], have been primarily developed for external, biologically non-interacting applications. The proprietary natures of these biomaterials retard the emergence of surface functionalization and elastic modulus tuning requisite for organs-on-chips. Therefore most current designs, like the lung-on-a-chip and gut-on-a-chip, rely on molding [35]. While this can be done with indirect printing, it is labor intensive and consumes more resources than a direct bioprinting approach.

2.2 Fabrication

State of the art organs-on-chips use photolithography to print negative molds, using SU-8 photoresist for example, and then form the flexible cellular substrate out of PDMS or similar elastomers [35]. These photopolymers are prime candidates for stereolithography (SLA) 3D printing. Rather than spin coating a silicon wafer and using a photomask to cure the SU-8, a process requiring seventeen individual steps in the lung-on-a-chip and the gut-on-a-chip [35], SLA 3D printing can automate the entire process. Using the Formlabs Form 1+SLA 3D printer as an example [41], the negative mold is designed with computer-aided design (CAD) software to generate a stereolithography format (STL) file. This STL file is then sliced to generate G-code, which provides instructions for the Form 1+. The printer cures the photopolymer in the desired 3D configuration, which is then used as the negative mold. A critical technical hurdle in this scenario is the layer thickness: the Form 1+has a lower limit of 25 μm [41] but some organ-on-a-chip structures need to be as thin as 2 μm [42]. Significant increases in print resolutions are required to make this a viable approach for all components.

While this approach can automate the design-to-mold process for thicker structures, there are still many steps remaining to produce a functioning organ-on-a-chip. The next advance will be to directly print elastomer structures that serve as the microfluidic channels, pneumatic actuators, and cell substrates. A shift away from PDMS may be necessary here because there is no current printing technology for this biomaterial. Even if it could be printed, there may be compromises on its elastic modulus or surface properties. There are new classes of biocompatible elastomers on the horizon, however, many of which could offer desirable physical properties. Some may be more tunable, have lower permeability to small molecules, or even allow for embedded electronics [43], which could serve as strain sensors or provide signals to nerve and muscle cells.

Once the structure is fabricated, either using 3D printing or more conventional methods, the next step is functionalization of the surface. This allows for adjustment of hydrophobicity, cell adhesion, chemical gradients, and perhaps even permeability of the biomaterial. Typically oxygen plasma is used to increase wetting of PDMS and silanization can restore hydrophobicity. Rajendra et al. demonstrate that low-cost inkjet printers can handle this task for paper-based microfluidic devices [44]. Similar thin layer printing could be used to modify the surfaces of elastomers in

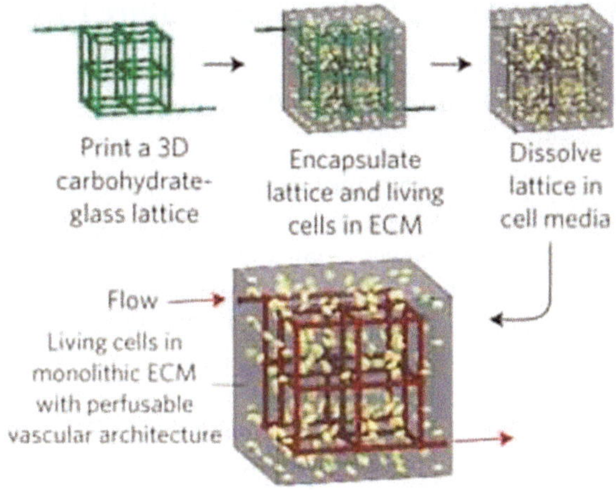

Fig. 4 Strategy for printing a perfusable vascular network within a block of ECM. This enables the construction of devices comprised of many cell layers. Figure is adapted from [45]

organs-on-chips. Deposition of ECM proteins can guide cell adhesion, while printing soluble morphogens can control differentiation and migration. If an elastomer structure has undesirable permeability characteristics, printing a sealant or permeabilizing agent could improve device function.

These speculative fabrication techniques only aim to leverage bioprinting methods to manufacture current organs-on-chips designs. Ultimately, in order to truly mimic the human body, degradable scaffold materials must be used so that the entire functional subunit of the device is fully biological. There are no published MPSs that eliminate permanent biomaterial substrates, but there has been significant progress in the 3D printing of biodegradable polymer scaffolds for tissue engineering applications. Sucrose lattices have been printed with a modified RepRap Mendel, which are then dissolved away to yield channels that can act as perfusable vascular networks within an ECM block for delivering nutrients to layered cells (Fig. 4) or can serve other microfluidic functions [45]. Printed molding of rigid scaffold materials, such as poly(D, L-lactic-co-glycolic acid) (PLGA), has been effective at replicating intestinal villi microstructures [46]. But it has not been demonstrated that these engineered tissues have the same mechanical properties and biological functions as their in vivo counterparts—the ability to withstand peristalsis and absorb drugs in this case.

Perhaps the most promising fabrication approach is to use bioprinted hydrogels. These biomaterials are well-characterized for their ability to serve as a temporary scaffold for cells. The hydrogels then degrade as the cells proliferate and produce their own sustainable ECM. Hydrogels are amenable to extrusion-based robocasting, which are 3D printing techniques that use syringes to deposit the biomaterial in layers. This approach has been used to create functional aortic valves [47], liver lobules [48], and reinforced cartilage [49] tissues with accurate mechanics and anatomy.

3 Regulatory Pathway and Future Directions

3.1 Regulatory Pathway

Organs-on-chips hold much promise for accelerating drug development by improving preclinical data quality and reducing study times and costs. In order to realize this promise, however, regulatory bodies must accept the data they generate. In the United States, the FDA is the gatekeeper to new pharmaceuticals, a role they execute in accordance with 21 CFR [39]. Given the myriad assays historically employed in IND-enabling preclinical studies, the FDA has quite a bit of discretion in what data to accept. The biggest hurdle here will be to take animal models out of the equation. The requirement for data from at least two models, including one non-rodent, is codified and can only be overcome with a revision of the CFR.

In order to convince the FDA and lawmakers of the utility of organs-on-chips, it may be best to use the devices to studying exiting compounds. By looking at both marketed pharmaceuticals and compounds that failed clinically, the predictive value of the devices can be established. Such studies should be done in accordance with GLP, which means that their fabrication will be subject to intense scrutiny. Standardization of bioprinting for this application will therefore be essential.

The future of personalized medicine will be enabled, in part, by MPSs like organs-on-chips. An increase in the overall rate of drug approvals coupled with lower development costs will give physicians much more discretion in prescribing treatments. Their decisions will rely on MPS-based diagnostics, assuming that the FDA consents. While there is much leeway in the law, there may not be enough to flexibility to accommodate the variability in organs-on-chips without additional legislation. This is similar to the challenges facing tissue engineering for regenerative medical applications [50]. By coupling the organ-on-a-chip diagnostic with the pharmaceutical development process, much like how imatinib was coupled with a genetic test for bcr-abl [1], researchers may be able to generate data that will usher in a change in the regulatory environment.

3.2 Future Directions and Concluding Thoughts

For all the promise of organ-on-chips, they are still in their infancy. The need for these chips to represent all of the major body tissues is clear: this is the only way to produce an accurate biomimetic human-on-a-chip. These integrated systems will have to be as diverse as humanity itself in order to predict pharmaceutical actions in various cohorts and offer insight into individual cases.

Bioprinting fabrication has the potential to advance this field by enabling new designs and customizations. This will rely heavily on the development of suitable biomaterials as well as the engineering of higher resolution 3D print technologies. The future of organs-on-chips appears most promising when the devices recapitulate the biology of parenchymal tissues fully. This means a move toward degradable and even self-assembling hydrogel scaffolds [51, 52].

When integrated systems are seeded with genetically diverse stem cells that are allowed to migrate and differentiate, the fabrication of such humans-on-chips may look much like the embryonic development of actual human bodies, just on a smaller scale. There inevitably will be ethical challenges that arise from such a scenario, but this can and must be balanced with the predictive value of these devices so that human suffering from disease can be alleviated.

Acknowledgements Financial support for this work was provided by the United States Department of Education, Washington, DC, and Harvard University, Cambridge, MA, USA. The authors indicate no potential conflict of interest.

References

1. Druker B, Tamura S, Buchdunger E, Ohno S, Segal G, Fanning S, et al. Effects of a selective inhibitor of the Abl tyrosine kinase on the growth of Bcr-Abl positive cells. Nat Med. 1996;2(5):561–6.
2. Herper M. The truly staggering cost of inventing new drugs. 2012. http://www.forbes.com/sites/matthewherper/2012/02/10/the-truly-staggering-cost-of-inventing-new-drugs/. Accessed 23 March 2015.
3. Willmann J, van Bruggen N, Dinkelborg L, Gambhir S. Molecular imaging in drug development. Nat Rev Drug Discov. 2008;7(7):591–607.
4. Lipinski C, Lombardo F, Dominy B, Feeney P. Experimental and computational approaches to estimate solubility and permeability in drug discovery and development settings. Adv Drug Deliv Rev. 2001;46(1–3):3–26.
5. U.S. Food and Drug Administration. Guidance for industry: M3(R2) nonclinical safety studies for the conduct of human clinical trials and marketing authorization for pharmaceuticals. Silver Spring; 2010. Accessed 23 March 2015.
6. Hay M, Thomas D, Craighead J, Economides C, Rosenthal J. Clinical development success rates for investigational drugs. Nat Biotechnol. 2014;32(1):40–51.
7. Paul S, Mytelka D, Dunwiddie C, Persinger C, Munos B, Lindborg S, et al. How to improve R & D productivity: the pharmaceutical industry's grand challenge. Nat Rev Drug Discov. 2010;9(3):203–14.
8. Lucey B, Nelson-Rees W, Hutchins G. Henrietta Lacks, HeLa cells, and cell culture contamination. Arch Pathol Lab Med. 2009;133(9):1463–7.
9. Egan W, Lauri G. Prediction of intestinal permeability. Adv Drug Deliv Rev. 2002;54(3):273–89.
10. Lennernas H, Palm K, Fagerholm U, Artursson P. Comparison between active and passive drug transport in human intestinal epithelial (caco-2) cells in vitro and human jejunum in vivo. Int J Pharm. 1996;127(1):103–7.
11. Sun D, Lennernas H, Welage L, Barnett J, Landowski C, Foster D, et al. Comparison of human duodenum and Caco-2 gene expression profiles for 12,000 gene sequences tags and correlation with permeability of 26 drugs. Pharm Res. 2002;19(10):1400–16.
12. Pampaloni F, Reynaud EG, Stelzer EHK. The third dimension bridges the gap between cell culture and live tissue. Nat Rev Mol Cell Biol. 2007;8:839–45.
13. National Human Genome Research Institute. Why mouse matters. 2010. http://www.genome.gov/10001345. Accessed 23 March 2015.

14. Pennisi E. ENCODE project writes eulogy for junk DNA. Science. 2012;337(6099):1159, 1161.
15. Huh D, Hamilton GA, Ingber DE. From 3D cell culture to organs-on-chips. Trends Cell Biol. 2011;21(12):745–54.
16. Singhvi R, Kumar A, Lopez GP, Stephanopoulos GN, Wang DI, Whitesides GM, et al. Engineering cell shape and function. Science. 1994;264(5159):696–8.
17. Dike LE, Chen CS, Mrksich M, Tien J, Whitesides GM, Ingber DE. Geometric control of switching between growth, apoptosis, and differentiation during angiogenesis using micropatterned substrates. Vitro Cell Dev Biol Anim. 1999;35(8):441–8.
18. Polte TR, Eichler GS, Wang N, Ingber DE. Extracellular matrix controls myosin light chain phosphorylation and cell contractility through modulation of cell shape and cytoskeletal prestress. Am J Physiol Cell Physiol. 2004;286(3):C518–28.
19. Lee PJ, Hung PJ, Lee LP. An artificial liver sinusoid with a microfluidic endothelial-like barrier for primary hepatocyte culture. Biotechnol Bioeng. 2007;97(5):1340–6.
20. Torisawa YS, Spina CS, Mammoto T, Mammoto A, Weaver JC, Tat T, et al. Bone marrow-on-a-chip replicates hematopoietic niche physiology in vitro. Nat Methods. 2014;11(6):663–9.
21. Jang KJ, Mehr AP, Hamilton GA, McPartlin LA, Chung S, Suh KY, et al. Human kidney proximal tubule-on-a-chip for drug transport and nephrotoxicity assessment. Integr Biol (Camb). 2013;5(9):1119–29.
22. Huh D, Matthews BD, Mammoto A, Montoya-Zavala M, Hsin HY, Ingber DE. Reconstituting organ-level lung functions on a chip. Science. 2010;328(5986):1662–8.
23. Kim HJ, Huh D, Hamilton G, Ingber DE. Human gut-on-a-chip inhabited by microbial flora that experiences intestinal peristalsis-like motions and flow. Lab Chip. 2012;12(12):2165–74.
24. Agarwal A, Goss JA, Cho A, McCain ML, Parker KK. Microfluidic heart on a chip for higher throughput pharmacological studies. Lab Chip. 2013;13(18):3599–608.
25. Galilei G. Dialogues concerning two new sciences by Galileo Galilei. New York: Macmillan; 1914.
26. Ng JM, Gitlin I, Stroock AD, Whitesides GM. Components for integrated poly(dimethylsiloxane) microfluidic systems. Electrophoresis. 2002;23(20):3461–73.
27. Palchesko RN, Zhang L, Sun Y, Feinberg AW. Development of polydimethylsiloxane substrates with tunable elastic modulus to study cell mechanobiology in muscle and nerve. PLoS One. 2012;7(12):e51499.
28. Brown XQ, Ookawa K, Wong JY. Evaluation of polydimethylsiloxane scaffolds with physiologically-relevant elastic moduli: interplay of substrate mechanics and surface chemistry effects on vascular smooth muscle cell response. Biomaterials. 2005;26(16):3123–9.
29. Zhou J, Khodakov DA, Ellis AV, Voelcker NH. Surface modification for PDMS-based microfluidic devices. Electrophoresis. 2010;33(1):89–104.
30. Wang L, Sun B, Ziemer KS, Barabino GA, Carrier RL. Chemical and physical modifications to poly(dimethylsiloxane) surfaces affect adhesion of Caco-2 cells. J Biomed Mater Res A. 2010;93(4):1260–71.
31. Moustafa ME, Gadepalli VS, Elmak AA, Lee W, Rao RR, Yadavalli VK. Large area micropatterning of cells on polydimethylsiloxane surfaces. J Biol Eng. 2014;8(1):24.
32. Hou S, Yang K, Qin M, Feng XZ, Guan L, Yang Y, et al. Patterning of cells on functionalized poly(dimethylsiloxane) surface prepared by hydrophobin and collagen modification. Biosens Bioelectron. 2008;24(4):918–22.
33. Patrito N, McCague C, Norton PR, Petersen NO. Spatially controlled cell adhesion via micropatterned surface modification of poly(dimethylsiloxane). Langmuir. 2007;23(2):715–9.
34. Ingber D. Synthetic human organs on chips. In ArtScience @Le Lab Seminar Series; 2014; Cambridge.
35. Huh D, Kim HJ, Fraser JP, Shea DE, Khan M, Bahinski A, et al. Microfabrication of human organs-on-chips. Nat Protoc. 2013;8(11):2135–57.
36. Duffy DC, McDonald JC, Schueller OJ, Whitesides GM. Rapid prototyping of microfluidic systems in poly(dimethylsiloxane). Anal Chem. 1998;70:4974–84.

37. Green PS. 3-D Printing's Promise—and Limits. 2014. http://www.wsj.com/articles/3-d-printings-promiseand-limits-1401473711. Accessed 23 March 2015.
38. Jonigk D, Golpon H, Bockmeyer CL, Maegel L, Hoeper MM, Gottlieb J, et al. Plexiform lesions in pulmonary arterial hypertension: composition, architecture, and microenvironment. Am J Pathol. 2011;179(1):167–79.
39. U.S. Food and Drug Administration. CFR—code of federal regulations Title 21. 2014. http://www.accessdata.fda.gov/scripts/cdrh/cfdocs/cfcfr/CFRsearch.cfm?CFRPart=58. Accessed 23 March 2015.
40. Stratasys. 3D Printing With Bio-compatible Material. 2015. http://www.stratasys.com/materials/polyjet/bio-compatible. Accessed 23 March 2015.
41. Forlabs. Form 1 + High-Resolution 3D Printer. 2015. http://formlabs.com/products/form-1-plus/. Accessed 23 March 2015.
42. Mathur A, Loskill P, Shao K, Huebsch N, Hong S, Marcus SG, et al. Human iPSC-based cardiac microphysiological system for drug screening applications. Sci Rep. 2015;9(5):8883.
43. Muth JT, Vogt DM, Truby RL, Mengüç Y, Kolesky DB, Wood RJ, et al. Embedded 3D printing of strain sensors within highly stretchable elastomers. Adv Mater. 2014;26(36):6307–12.
44. Rajendra V, Sicard C, Brennan JD, Brook MA. Printing silicone-based hydrophobic barriers on paper for microfluidic assays using low-cost ink jet printers. Analyst. 2014;139(24):6361–5.
45. Miller JS, Stevens KR, Yang MT, Baker BM, Nguyen DHT, Cohen DM, et al. Rapid casting of patterned vascular networks for perfusable engineered three-dimensional tissues. Nat Mater. 2012;11(9):768–74.
46. Lee M, Dunn JCY, Wu BM. Scaffold fabrication by indirect three-dimensional printing. Biomaterials. 2005;26(20):4281–9.
47. Hockaday LA, Kang KH, Colangelo NW, Cheung PYC, Duan B, Malone E, et al. Rapid 3D printing of anatomically accurate and mechanically heterogeneous aortic valve hydrogel scaffolds. Biofabrication. 2012;4(3):035005.
48. Gou M, Qu X, Zhu W, Xiang M, Yang J, Zhang K, et al. Bio-inspired detoxification using 3D-printed hydrogel nanocomposites. Nat Commun. 2014;5:3774.
49. Bakarich SE, Gorkin R, Panhuis Mih, Spinks GM. Three-dimensional printing fiber reinforced hydrogel composites. ACS Appl Mater Interfaces. 2014;6(18):15998–16006.
50. Lloyd-Evans M. Regulating tissue engineering. Mater Today. 2004;7(5):48–55.
51. Qi H, Ghodousi M, Du Y, Grun C, Bae H, Yin P, et al. DNA-directed self-assembly of shape-controlled hydrogels. Nat Commun. 2013;4:2275.
52. Appel EA, Tibbitt MW, Webber MJ, Mattix BA, Veiseh O, Langer R. Self-assembled hydrogels utilizing polymer–nanoparticle interactions. Nat Commun. 2015;6:6295.

Index

K. Turksen (ed.), *Bioprinting in Regenerative Medicine,* Stem Cell Biology and Regenerative Medicine, DOI 10.1007/978-3-319-21386-6

MIX
Papier aus verantwortungsvollen Quellen
Paper from responsible sources
FSC® C105338

If you have any concerns about our products,
you can contact us on
ProductSafety@springernature.com

In case Publisher is established outside the EU,
the EU authorized representative is:
Springer Nature Customer Service Center GmbH
Europaplatz 3, 69115 Heidelberg, Germany

Printed by Libri Plureos GmbH
in Hamburg, Germany